Student Study Guide

to accompany

Concepts in Biology

Ninth Edition

Eldon D. Enger
Delta College

Frederick C. Ross
Delta College

Boston   Burr Ridge, IL   Dubuque, IA   Madison, WI   New York   San Francisco   St. Louis
Bangkok   Bogotá   Caracas   Lisbon   London   Madrid
Mexico City   Milan   New Delhi   Seoul   Singapore   Sydney   Taipei   Toronto

McGraw-Hill Higher Education

*A Division of The **McGraw-Hill** Companies*

Student Study Guide to accompany
CONCEPTS IN BIOLOGY, NINTH EDITION

Copyright ©2000 by The McGraw-Hill Companies, Inc. All rights reserved.
Previous editions copyright year, © 1996, 1993, 1990.
Printed in the United States of America.

The contents of, or parts thereof, may be reproduced for use with
CONCEPTS IN BIOLOGY by Eldon D. Enger and Frederick C. Ross,
provided such reproductions bear copyright notice and may not be reproduced
in any form for any other purpose without permission of the publisher.

RECYCLED

This book is printed on recycled, acid-free paper
containing 10% postconsumer waste.

3 4 5 6 7 8 9 0 QPD/QPD 9 0 3 2 1 0 9

ISBN 0-697-36056-3

www.mhhe.com

Contents

Preface

To the Student

This student study guide is an aid to learning the material covered in *Concepts in Biology, 9th ed.* There is no one best way to study biology. You must discover the best combination of techniques that work for you. This study guide is designed to allow you to choose from several study aids, depending on how you learn best. You will note the emphasis on vocabulary. Frequently students find the language of science to be intimidating. However, rather than allowing the terminology to become a stumbling block, learn a few new terms each time you study. You will find that the quantity of new works is quite manageable if taken in small bits. The study of linguistics (the Sapir-Whorf Hypothesis) teaches that language shapes thought. People think in the words and related concepts our language gives us. It is easier to conceive of something if we have a word for it. Words or terminology help give substance to ideas.

The organization of the study guide is related to the organization of the text, *Concepts in Biology, 9th ed.* There is a section of the study guide related to each chapter in the text. In each section of the study guide there are several features that are consistent. *Page 1* of the study guide for each chapter has a statement that provides an overview of the chapter followed by a list of study activities.

Next, there is space entitled "Key Terms/Notes." This is a chronological listing of most of the **boldface** terms used in the chapter. The first terms used in the chapter should be at the top of the list and so on. Use the space provided to make notes on the terms that help you remember and understand their meaning. Perhaps you want to note the page number where the term is first described. Perhaps you want to indicate how your instructor modified or expanded on the term's usage. You may just want to write the definition for yourself in your own words. Use this space as a study too. Let it help you accumulate the language of biology. The more you write, the more you will learn. As one student commented, "If it comes out the end of my pencil, it sticks in my head."

There are three kinds of questions to help you measure whether you have met the objectives or not. "Questions with Short Answers" are fill-in-the-blank-type questions that are straightforward descriptions of terms or processes with essential parts of the statement left blank. After reading and studying the chapter, read these statements and put your answer on the blank lines provided. You can then check your answers in the "Student Study Guide Answers" section at the back of this study guide. Note how the word or words make the statement a complete thought. Check your answer. Does it fit as well or better than the answer provided? Does your answer have implications that the answer given in the study guide does not? If you have not chosen exactly the same set of words to fill in the blank space, make this a learning situation. Be sure that you do not just check to see if your answer is correct, but let the answer given be a study tool.

A second set of questions is identified as "Label/Diagram/Explain." This is an activity that gives you directions and lets you explore several different ways of organizing the answer. The more you think about the answer the more deeply you will understand the material in the chapter, and the more details you may include. At the end of the study guide there is a sample answer to each of these questions. After you have created your response, look at the one provided. Does it include each of the details you have included? Is it more detailed than your answer? Compare the answers and make sure you understand the importance of the different ways of expressing the answer. Remember that this should be a learning tool.

Finally, there are a series of multiple choice questions in each section of this study guide. These questions ask you to think about the information you have learned. They require you to use combinations of information you may have acquired from the text, lab, lecture, or other resources. They will require you to evaluate several possible answers, only one of which is the best answer. Some responses are obviously wrong. Others may be partially correct, and therefore misleading. It is up to you to pick apart each response to determine why it may be *incorrect*. Select the answer you think is most correct for each question and indicate it beside the number of the question. Compare your answers with the answers in the back of this study guide. If your choice was different from that given, be sure to reread the question to make sure you completely understand why one response is better than the other. Be aware, three of the four responses are *wrong*. Spend time understanding why these are wrong and you will also be learning the material. Concentrating only on learning the right answer is only one fourth of the material you need to know.

The most important thing to remember about a student study guide is that it is a tool that should help you learn the material. Use as much of the study guide as will help. We would appreciate your comments about whether or not this was a helpful tool for you.

To the Instructor

Each student learns differently. The study guide provides several options for your students to enable them to learn more effectively. Some students may need to have the study guide assigned, others may take hints and suggestions and utilize what is valuable to them and leave the rest. Several levels of questions are provided: rote memory aids (fill-in-the-blank and vocabulary) and higher order learning (multiple choice and explanation).

We welcome your comments concerning the value of the student study guide and vocabulary deck.

CHAPTER 1
WHAT IS BIOLOGY?

Overview

A logical place to start is with a general introduction to the nature of science and the significance of biology in your everyday life. Here we present a scientist's view of the world and describe what living things are and how they differ from nonliving things. The purpose is to lay the groundwork for helping you understand and answer questions about living things in your environment. There may be a number of different answers for each question, yet the answers may not be simple. You will be better able to understand and answer biological questions after you have an understanding of how science works.

Study Activities

1. Write a summary of each section of the *Chapter Outline* in your text.
2. For each of the *Learning Objectives* in your text, write a sentence or paragraph that demonstrates your mastery of the objective.
3. Answer the *Questions* at the end of the chapter in your text.
4. Complete the student study guide.

Key Terms/Notes

Define each of the following terms in the space provided, or make flash cards of the following terms.

Biology

Science

Scientific method

Valid

Reliable

Empirical evidence

Hypothesis

Controlled experiment

Control group

Experimental group

Variable

Theory

Scientific law	Control processes
Theoretical science	Enzymes
Applied science	Organism
Metabolic processes	Tissue
Generative processes	Organ
Responsive processes	Organ system

Questions with Short Answers

1. The science that deals with things which are living is _____.

2. Astrology, music, and literature are examples of_____.

3. An organism responds to its environment, it changes from season to season, and over time the population changes. These are _____ processes.

4. Organisms making more of themselves is an example of a _____.

5. All living things are composed of similar basic units of life, which are called _____.

6. Metabolic processes include such things as nutrient _____, nutrient processing, and waste elimination.

7. A group of similar cells which cooperate to perform a particular job are known as a _____.

8. Applied sciences attempt to find a practical solution to a particular problem, whereas _____ science is an attempt to gain new knowledge for its own sake.

9. A logical guess at a correct answer to a question or solution to a problem is generally known as a(n)_____.

10. Information that is gained by observation is called _____ evidence.

Label/Diagram/Explain

Here is a scenario. Put this information into a scientific perspective by identifying each step of the Scientific Method and indicating what the scientific team did as they performed each step.

Researchers in one scientific laboratory have noticed that the plants they have been studying seem to become tall and spindly as a result of low light levels. One of their members has read a report of "legginess" in plants under low light. They decide to determine if high levels of light are required for the plants to grow "normally." They establish that normal plants grow up to 8 centimeters with attractively spaced leaves and design an experiment to verify if plants grow taller with greater spacing between the leaves with lower light levels. When they published their data another scientific team found that they could not get their plants to react in exactly the same way all the time, but 9 out of 10 plants did behave in a similar way, under similar conditions. The first scientific team put one group of 50 cuttings from a plant into an area that had 1/2 as much light as in a greenhouse. They placed 50 other cuttings in a greenhouse. After six weeks, the comparison was stopped, but in the meantime, height measurements were taken three times each week, and recorded in a table. A mathematical analysis of their data determined that it was significant.

Observation:

Control group:

Question formulation:

Experimental group:

Explore alternative resources:

Publication and peer review:

Hypothesis:

Multiple Choice Questions

1. Which one of the following is NOT an example of how biological information is used?
   A. development of catalytic converters
   B. disposal methods for infectious waste
   C. using bacteria to produce medicines for humans
   D. collecting data on clear cutting of forested areas

2. Science can be differentiated from nonscience because scientific results are
   A. repeatable
   B. always collected in a laboratory
   C. based on single events
   D. formed from opinions

3. Which one of the following represents a generative process?
   A. enzymes
   B. individual adaptation
   C. nutrient uptake
   D. cell division

4. Which of the following is most true?
   A. the progress of science is determined by the kinds of questions asked
   B. a stated hypothesis is always correct
   C. scientists cannot make errors
   D. empirical evidence will not support hypotheses

5. Which of these is a biological problem?
   A. famine in Rhodesia
   B. hurricanes killed thousands in Honduras
   C. pesticide research leads to toxic waste
   D. all of these have a biological component

6. When dealing with responsive processes:
   A. metabolic processes decrease.
   B. populations evolve through time.
   C. organisms grow.
   D. individuals coordinate activities.

7. Scientific conclusions are reliable if
   A. they are supported by empirical evidence.
   B. the same results are obtained in successive trials.
   C. the scientific method was followed.
   D. they support the hypothesis.

8. Empirical evidence
   A. is information gained by observation.
   B. may be obtained directly by the senses or indirectly with instruments.
   C. should be verified or disproved by further observation.
   D. all of the above.

9. Which of the following has the least supporting evidence?
   A. theory
   B. law
   C. hypothesis
   D. all of the above have equal supporting evidence

10. Which of the following is the most specific?
    A. theory
    B. law
    C. hypothesis
    D. all of the above are equally specific

11. A plausible, scientifically acceptable generalization is
    A. a theory.
    B. a law.
    C. a hypothesis.
    D. empirical evidence.

12. Which of the following is an applied science?
    A. particle physics
    B. evolution
    C. astronomy
    D. genetic engineering

13. In a controlled experiment, the experimental group
    A. should be identical to the control group.
    B. is not necessary if there is a control group.
    C. differs from the control group by one variable.
    D. should consist of one test individual.

14. Metabolic processes include
    A. nutrient processing.
    B. waste elimination.
    C. nutrient uptake.
    D. all of these.

15. Which is *least* likely to be understood using the scientific method?
    A. if it is possible to attack someone by their smell
    B. if life ever existed on Mars
    C. how to float your boat
    D. if you will be successful in this course

CHAPTER 2
SIMPLE THINGS OF LIFE

Overview
In order to understand the structure and activities of living organisms, you must know something about the materials from which they are made. Here we discuss the structure of matter and the energy it contains. As you study this material, you should consciously try to build a vocabulary that will help you describe matter.

Study Activities
1. Write a summary of each section of the *Chapter Outline* in your text.
2. For each of the *Learning Objectives* in your text, write a sentence or paragraph that demonstrates your mastery of the objective.
3. Answer the *Questions* at the end of the chapter in your text.
4. Complete the student study guide.

Key Terms/Notes
Define each of the following terms in the space provided, or make flash cards of the following terms.

Matter

Electron

Element

Atomic number

Atom

pH

Chemical symbol

Atomic mass unit

Atomic nucleus

Isotope

Proton

Periodic table of the elements

Neutron

Energy level

Ions	Diffusion
Chemical bonds	Chemical Reactions
Acid	Oxidation-reduction
Molecule	Reactant
Hydrogen bond	Product
Kinetic energy	Mixture
State of matter	Cation/Anion

Questions with Short Answers

1. The basic building block of all things is a unit called a(n) _____.

2. A molecule moves faster or slower depending on the amount of _____ it contains.

3. Atoms are composed of these three parts, _____, neutrons, and electrons.

4. The protons are located in the _____ of the atom.

5. _____ spend their time at some distance from the nucleus of the atom.

6. Neutrons have a _____ charge.

7. An isotope is an atom which has a specific number of _____ in its nucleus.

8. Atoms that share electrons form what kind of bond? _____

9. Ionic bonds are formed between two _____.

10. One bond that does not form a molecule is known as a(n)_____ bond.

11. Three states of matter are_____, liquid, and gas.

12. A_____ has more energy than a liquid.

13. The one type of mixture that will settle out is a_____.

14. Rearranging atoms within a molecule result from a chemical_____.

15. Unstable isotopes which spontaneously disintegrate are said to be_____.

16. The electrons which have the greatest_____ are located farthest from the nucleus.

17. A negative ion is one which has an additional_____.

18. The scale upon which the relative amounts of acidity is measured in the_____scale.

19. A_____ is a group of atoms bonded to each other in specific proportions.

20. _____ are molecules that have an unequal distribution of charges on their surface; ex. water.

Label/Diagram/Explain

Potassium (K) is atomic number 19 on the periodic table of the elements. One isotope of potassium contains 21 neutrons. Answer the following questions about this isotope:

1. What is its atomic mass?_____

2. Where are the neutrons and protons located in this atom, and how many of each are present?_____

3. How many electrons are present in this isotope of potassium?_____

4. When this isotope forms an ion, the ion will have_____ electrons.

5. What is the charge of the ion formed?_____

6. In a chloride ion, there will be_____ in its outer most energy level.

Multiple Choice Questions

1. How many ions of hydrogen are able to bond to one ion of sulfur?
   A. only one
   B. two ions of hydrogen
   C. three hydrogen ions
   D. four or more ions

2. An isotope has a specific number of neutrons, whereas an ion has a(n):
   A. equal number of neutrons/electrons.
   B. as many protons as neutrons.
   C. only protons and electrons.
   D. specific number of electrons.

3. A neutron is an example of a:
   A. compound.
   B. sub-atomic particle.
   C. mixture.
   D. hydrogen bonded solution.

4. When a liquid is heated:
   A. it takes up more space.
   B. its state changes to a solid.
   C. energy is absorbed when it is cooled.
   D. it has more density than before.

5. Which electrons have the most energy?
   A. those closest to the nucleus
   B. electrons in pairs
   C. an electron in the outermost energy level
   D. no correct answer, all electrons are equal

6. "Useful energy is lost during reactions" is part of the:
   A. Universal Chaos Principle.
   B. Second Law of Thermodynamics.
   C. Interdependence of Matter/Energy.
   D. First Law of Thermodynamics.

7. Diffusion, the net movement of molecules is the result of:
   A. flow of electrons from one atom to another atom.
   B. the kinetic energy of the molecules.
   C. rearrangement of bonds.
   D. an activity of only living things.

8. A characteristic of a radioactive isotope is:
   A. formation of high energy bonds.
   B. difficulty/inability to react.
   C. loss of its electrons.
   D. instability of its nucleus.

9. A solution with a pH of 3
   A. has a high concentration of hydroxyl ions.
   B. is a base.
   C. has a high concentration of hydrogen ions.
   D. is neutral.

10. Aluminum has an atomic number of 13 and an atomic mass number of 26.98. How many neutrons are in a typical atom of aluminum?
    A. 13
    B. 14
    C. 26
    D. 27

11. An ion with 10 electrons, 11 protons, and 12 neutrons will have a charge of
    A. +.
    B. −.
    C. ++.
    D. − −.

12. Compared to ^{12}C, the isotope ^{14}C has
    A. a different atomic number.
    B. two more neutrons.
    C. two more protons.
    D. two more electrons.

13. In the reaction $HCl + NaOH \longrightarrow NaCl + H_2O$
    A. an acid is produced.
    B. a base is produced.
    C. a salt is produced.
    D. all of the above.

14. The orientation of adjacent polar water molecules is such that
    A. the oxygen end of one water molecule is attracted to the oxygen end of the other water molecule.
    B. a hydrogen of one water molecule is attracted to the oxygen end of the other water molecule.
    C. a hydrogen of one water molecule is attracted to a hydrogen of the other water molecule.
    D. the two molecules repel each other.

15. An atom with 2 electrons in its outer energy level will likely
    A. form hydrogen bonds with other atoms.
    B. share 2 electrons with other atoms.
    C. lose 2 electrons and become anions.
    D. lose 2 electrons and become cations.

CHAPTER 3
ORGANIC CHEMISTRY: THE CHEMISTRY OF LIFE

Overview

The chemistry of living things is really the chemistry of the carbon atom and a few other atoms that can combine with carbon. In order to understand some aspects of the structure and function of living things, you should first learn some basic organic chemistry. It is the intent of this chapter to provide this background.

Study Activities

1. Write a summary of each section of the *Chapter Outline* in your text.
2. For each of the *Learning Objectives* in your text, write a sentence or paragraph that demonstrates your mastery of the objective.
3. Answer the *Questions* at the end of the chapter in your text.
4. Complete the student study guide.

Key Terms/Notes

Define each of the following terms in the space provided, or make flash cards of the following terms.

Organic molecules

Functional groups

Biochemistry

Dehydration synthesis

Covalent bond

Hydrolysis

Double bond

Carbohydrate

Empirical formula

Complex carbohydrates

Structural formula

Lipid

Carbon skeleton

Fats

Saturated fatty acid

Peptide bond

Poly-unsaturated

Polypeptide

Phospholipid

Denature

Steroid

Nucleotide

Protein

Nucleic acid

Amino acid

Monomer/Polymer

Questions with Short Answers

1. The characteristic atom contained in all organic molecules is _____.

2. An atom of carbon has _____ bondable electrons.

3. Each of the bondable electrons of a carbon atom can form a _____ bond.

4. Organic molecules that have the exact same numbers of a kind of atom may be very different in _____ and different in activity.

5. The group of carbon atoms that provide the framework for an organic molecule may be known as the

   _____.

6. Those atoms attached to the carbon skeleton are known as _____ groups.

7. The FOUR major groups of organic molecules include: _____, lipids, nucleic acids, and proteins.

8. Carbohydrates are composed of three kinds of atoms, carbon, hydrogen, and _____.

9. _____ have fewer hydrogen atoms than do lipids.

10. The building blocks of neutral fats are triple alcohols known as glycerol and _____.

11. If a fatty acid has no double bonds between the carbon atoms, it is known as a _____ fatty acid.

12. The building blocks (monomers) of a protein are called _____.

13. Amino acids are bonded together by _____ bonds to form polymers called proteins.

14. When two atoms share four electrons they form a _____ bond.

15. Nitrogen is able to form _____ covalent bonds.

16. A functional group which contains • OH is an _____.

17. A _____ is composed of several monomers attached by dehydration synthesis reactions.

18. A _____ is a single monomer of a complex carbohydrate.

19. There are about 20 different kinds of_____which are used to construct various proteins.

Label/Diagram/Explain

1. Draw the structural formulae for each of the following molecules:

CO_2 C_2H_4O

H_2O HCN

CH_3OH NH_3

2. Identify each of the functional groups below.

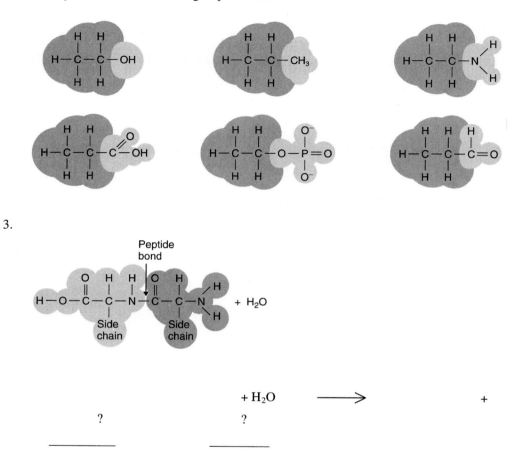

3.

What type of molecule is this? _____

Identify the (1) amino group and (2) carboxylic acid group on the molecule above.

Draw in the missing molecules for the reaction above.

What type of reaction does this represent? _____

4.

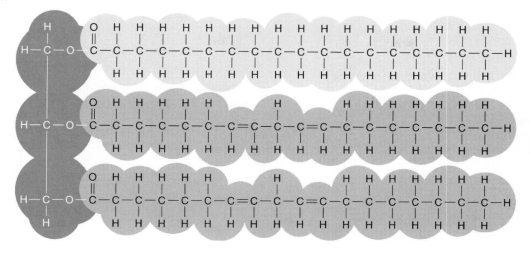

What type of molecule is this? _____

How is this molecule different than a phospholipid?

_____ _____

Circle the following parts of the molecule using different colored pencils or markers: (1) glycerol, (2) a saturated fatty acid and (3) unsaturated fatty acids on the molecule above.

5. Complete the following reaction. Identify the reactants. Identify the products. What type of reaction is this?

$$C_5H_{10}O_5 + C_5H_{10}O_5 \longrightarrow \quad ? \quad + \quad ?$$

_____ _____

Multiple Choice Questions

1. How many atoms of hydrogen are able to bond with one atom of carbon?
   A. only one
   B. two atoms of hydrogen
   C. three hydrogen atoms
   D. four atoms of hydrogen

2. An atom of nitrogen has a specific number of bondable electrons based on:
   A. their placement in orbitals.
   B. the number of neutrons in the nucleus.
   C. the size of the molecule.
   D. atomic mass of the atom.

3. Which of the following is organic?
   A. sulfuric acid
   B. copper sulfate
   C. saltwater solution
   D. lipid

4. Methyl, alcohol, and amino are all examples:
   A. functional groups.
   B. carbon skeletons.
   C. side chains of polysaccharides.
   D. places where hydrolysis can occur.

5. • COOH is a functional group known as:
   A. phosphate.
   B. carbonyl group.
   C. carboxylic acid group.
   D. amino.

6. Hydrolysis forms:
   A. small molecules from large ones.
   B. water as a waste product.
   C. rearranged molecules of the same size.
   D. condensed water molecules.

7. An enzyme's function is to:
   A. be a primary energy source.
   B. act as an organic catalyst.
   C. store energy for long term use.
   D. form high energy bonds.

8. Glucose ($C_6H_{12}O_6$), galactose ($C_6H_{12}O_6$), and fructose ($C_6H_{12}O_6$) are
   A. the building blocks of complex carbohydrates.
   B. isomers.
   C. hexose sugars.
   D. all of the above.

9. Which one of the following is NOT a lipid?
   A. neutral fat
   B. phospholipid
   C. polypeptide
   D. testosterone

10. Which of the following is NOT a function of neutral fat?
    A. insulation
    B. shock absorption
    C. energy storage
    D. regulate rates of chemical reactions

11. Which is the monomer for a nucleic acid?
    A. amino acid
    B. nucleotide
    C. fatty acid
    D. monosaccharide

12. Which of the following is a polymer?
    A. starch
    B. protein
    C. DNA
    D. all of the above

13. $C_{12}H_{22}O_{11} + H_2O \rightarrow C_6H_{12}O_6 + C_6H_{12}O_6$
    The reaction above is an example of

    A. dehydration synthesis.
    B. the synthesis of a fat molecule.
    C. hydrolysis.
    D. the synthesis of a disaccharide.

14. Peptide bonds
    A. result from dehydration synthesis reactions.
    B. form between amino acids.
    C. occur when a carboxylic acid group reacts with an amino group.
    D. all of the above.

15. Which association is correct?

Nucleic Acid	Protein	Lipid	Carbohydrate
A. RNA	enzymes	insulin	glucose
B. DNA	insulin	testosterone	cellulose
C. Enzymes	insulin	testosterone	glucose
D. RNA	enzymes	cellulose	glycerol

CHAPTER 4
CELL HISTORY, STRUCTURE, AND FUNCTION

Overview
The cell is the simplest structure capable of existing as an individual living unit. Within this unit, certain chemical reactions are required for maintaining life. These reactions do not occur at random, but are associated with specific parts of the many kinds of cells. You need to recognize certain cellular structures found within most types of cells and describe their functions.

Study Activities
1. Write a summary of each section of the *Chapter Outline* in your text.
2. For each of the *Learning Objectives* in your text, write a sentence or paragraph that demonstrates your mastery of the objective.
3. Answer the *Questions* at the end of the chapter in your text.
4. Complete the student study guide.

Key Terms/Notes
Define each of the following terms. Label the cell diagram that follows with the structures identified with an asterisk (*).

Microscope

Permeability

Protoplasm

Chromosome

Cellular membranes

Active transport

Diffusion

Pinocytosis

Osmosis

Endoplasmic reticulum

Net movement

Aerobic cellular respiration

Chloroplast

Chlorophyll

Photosynthesis

Phagocytosis

Concentration gradient

Facilitated diffusion

Cytoskeleton

Hydrophobic

Hydrophilic

Hypertonic

Hypotonic

Eukaryote

Prokaryote

Nucleoplasm

Cytoplasm

*Ribosome

*Microtubules

*Centriole

*Cilia/Flagella

*Golgi apparatus

*Lysosome   *Nucleolus

*Nuclear membrane   *Nucleus

*Mitochondrion   *Chromatin

*Vacuoles   *Smooth endoplasmic reticulum

*Plasma membrane   *Rough endoplasmic reticulum

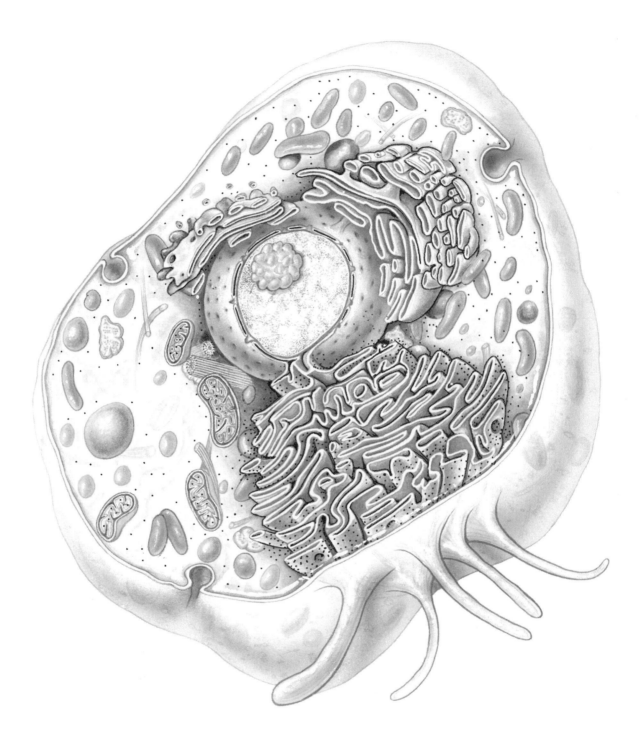

Questions with Short Answers

1. A cytoskeleton is composed of _____, microfilaments, and intermediate fibers.

2. Cell membranes are composed of two _____ layers and globules of protein.

3. The material inside of the nucleus is known as _____.

4. Movement of molecules from an area of higher concentration to an area of lower concentration is _____.

5. If a group of molecules is moved into a cell where there is a high concentration of these molecules the process may require an input of energy. This process is known as_____.

6. Those structures within a cell that perform particular functions are called _____.

7. Three cytoplasmic organelles formed from membranes are the endoplasmic reticulum, the _____ apparatus, and lysosomes.

8. Mitochondria are membranous cytoplasmic organelles where carbohydrate _____ takes place.

9. The green colored organelles in plant cells are the _____.

10. Photosynthesis is the process whereby plant cells make _____ using (sun) light as the source of energy.

11. Globules of RNA and protein located in the cytoplasm are known as _____.

12. The _____ is composed of microtubles and used at the time of cell division.

13. Flagella and _____ are both structures that may provide locomotion to a cell.

14. Cells which have numerous membranous organelles are likely to be classified as _____.

15. A _____ is a large membranous container within a cell.

Label/Diagram/Explain

1. The drawings below represent normal healthy blood cells. At the right, make drawings to illustrate how these cells will look when placed in hypotonic, hypertonic, and isotonic solutions. Use arrows to indicate the net direction of water movement across the plasma membrane. Also identify the areas where water concentration is highest.

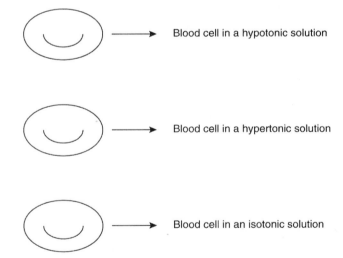

Blood cell in a hypotonic solution

Blood cell in a hypotonic solution

Blood cell in a hypertonic solution

Blood cell in a hypertonic solution

Blood cell in an isotonic solution

Blood cell in an isotonic solution

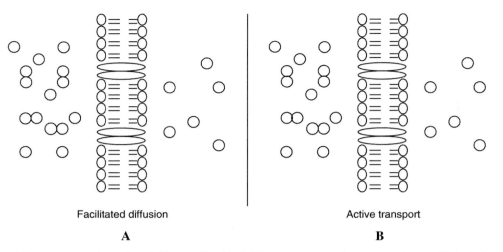

Facilitated diffusion   Active transport

A   B

2. The center portion of each illustration (A & B) represents a plasma membrane. The circles on either side of the membranes represent sugar molecules.
   a. Label the phospholipids and carrier protein molecules.
   b. Identify the hydrophobic and hydrophilic ends of the phospholipids.
   c. Draw arrows on the first figure to represent the direction facilitated diffusion will move the sugar molecules across the membrane.
   d. Draw arrows on the second figure to represent the direction active transport will move the sugar molecules across the membrane.
   e. Which process, facilitated diffusion or active transport, requires cellular energy? In what form does this energy occur?

Multiple Choice Questions

1. Long cellular organelles of movement are:
   A. cilia.
   B. microfilaments inside cell.
   C. flagella.
   D. endoplasmic reticulum.

2. Structures which are not well defined within the cytoplasm are called:
   A. microfilaments.
   B. cytoplasmic inclusions.
   C. protoplasm components.
   D. pinocytic vesicles.

3. Water diffuses into a cell by
   A. active transport.
   B. facilitated diffusion.
   C. phagocytosis.
   D. osmosis.

4. The organelle composed primarily of DNA:
   A. ribosome.
   B. nucleoplasm.
   C. nucleolus.
   D. chromosome.

5. In the phospholipid layer of a cell membrane the _____ ends of the phospholipids are on the outside.
   A. hydrophobic
   B. hydrophilic
   C. hypertonic
   D. hypotonic

6. The diffusion gradient is the difference between
   A. small and large particles.
   B. soluble vs. insoluble.
   C. high to low concentrations.
   D. outside and inside a cell.

7. A cell that is 97% water is placed in pure water.
   A. The cell is now in a hypertonic environment.
   B. Solute will diffuse out of the cell until a point of equilibrium is reached.
   C. The cell will shrink.
   D. Water will diffuse into the cell.

8. Which of the following moves molecules from an area of low concentration to an area of high concentration?
   A. osmosis
   B. facilitated diffusion
   C. active transport
   D. all of the above

9. White blood cells (leukocytes) engulf invading bacteria and viruses by
   A. pinocytosis.
   B. budding.
   C. active transport.
   D. phagocytosis.

10. Which one of the following is NOT composed of membrane?
    A. mitochondria
    B. centrioles
    C. chloroplasts
    D. golgi bodies

11. Hydrolytic enzymes and proteins called *defensins* move from the _____ into the vacuole to destroy the microorganisms.
    A. ribosomes
    B. peroxisomes
    C. mitochondria
    D. lysosomes

12. Which of the following is not a component of a prokaryotic cell?
    A. ribosomes
    B. chloroplast
    C. plasma membrane
    D. chromatin

13. The site of protein synthesis is the
    A. ribosome.
    B. golgi apparatus.
    C. nucleus.
    D. mitochondria.

14. Protein molecules are required for
    A. osmosis.
    B. active transport and facilitated diffusion.
    C. diffusion.
    D. concentration gradients.

CHAPTER 5
ENZYMES

Overview

Living cells require various chemical reactions to conduct their vital functions. To prevent the malfunction and death of the cell, these reactions must be conducted rapidly and be controlled. The problem is not starting reactions but controlling the rate of the reactions. Regulation of the rates of the many reactions in cells is the task of the enzymes. Enzymes function best when they have the proper 3-dimensional shape. Their shape is exquisitely sensitive to temperature and pH. A slight modification in temperature or pH can render the enzyme inactive. For this reason, the body has mechanisms which serve to keep both temperature and pH within a narrow range for enzyme operation.

Study Activities

1. Write a summary of each section of the *Chapter Outline* in your text.
2. For each of the *Learning Objectives* in your text, write a sentence or paragraph that demonstrates your mastery of the objective.
3. Answer the *Questions* at the end of the chapter in your text.
4. Complete the student study guide.

Key Terms/Notes

Define each of the following terms in the space provided, or make flash cards of the following terms.

Activation energy

Active site

Catalyst

Coenzyme

Enzyme

Turnover number

Substrate

Denature

Enzyme-substrate complex

Control processes

Attachment site

Enzymatic competition

Gene regulator proteins   Inhibitor

Negative feedback inhibition

Questions with Short Answers

1. The place on an enzyme where the substrate fits so that a reaction may occur is called the _____.

2. In order for a reaction to start there is energy necessary called _____ energy.

3. A _____ is a molecule (either organic or inorganic) that changes the rate of a reaction.

4. A protein that acts as a catalyst is termed a(n) _____.

5. A _____ aids in a reaction by removing one of the end products or by bringing in part of the substrate.

6. When an enzyme temporarily attaches to a reactant molecule it forms an _____.

7. For an enzyme to attach to a substrate it must be the correct _____.

8. Very large increases in temperature tend to _____ enzymes.

9. Changes in pH cause the enzyme to change _____.

10. The number of molecules of substrate which can be worked on in a given time is termed the enzyme's _____.

11. Increasing the temperature slightly might _____ the turnover number.

12. Negative feedback _____ controls the synthesis of product molecules by effecting enzyme action, not gene action.

13. Molecules called _____are chemical messengers that inform the genes as to whether protein production should occur.

14. When several enzymes are able to combine with a given substrate, _____ results.

15. Enzymes _____ the activation energy required for a reaction to occur.

16. The reactant molecule in an enzyme assisted reaction is called a _____.

17. Enzyme names frequently end in _____.

Label/Diagram/Explain

1. Label the diagram below. Your labels should include: active site, end products, enzyme, substrate, and enzyme-substrate complex. In the space below, explain how an enzyme works.

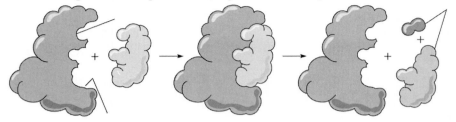

2. Complete the graph below to represent how an enzyme is able to lower activation energy. Explain the graph.

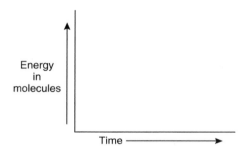

3. Complete the graph below to represent the effect of temperature on turnover number. Identify the optimum temperature and the temperatures that denature the enzyme. Describe in detail how temperature changes influence enzyme activity. Be sure that your answer includes information concerning the three-dimensional shape of the enzyme.

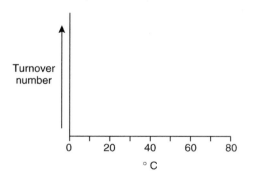

4. Complete the graph below to represent how changes in pH effect enzyme activity. Describe in detail how changes in pH influence enzyme activity.

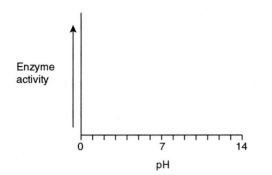

5. How will increases in concentration of either the enzyme or the substrate molecules influence the amount of product that is produced in a given amount of time? Explain.

6. Describe how an inhibitor works. How does the presence of inhibitor molecules influence enzyme activity?

Multiple Choice Questions

1. A vitamin sometimes works as a(n):
   A. enzyme.
   B. inhibitor.
   C. coenzyme.
   D. catalyst.

2. Competition between enzymes causes:
   A. enzymes to denature.
   B. decreases in the formation of certain products.
   C. each enzyme to function better.
   D. rearrangement of active sites.

3. As the temperature of an enzyme controlled reaction is moderately increased,
   A. changes in concentration of enzyme results.
   B. fewer transitory molecules are formed.
   C. decreases in turnover number occur.
   D. more collisions occur so more of them are effective.

4. Enzyme names frequently
   A. include the name of the substrate.
   B. ends in -ase.
   C. indicate the type of reaction they facilitate.
   D. all of the above.

5. Turnover number:
   A. increases as pH increases.
   B. does not change relative to concentration of acid or base.
   C. increases as optimum conditions are approached.
   D. is constant.

6. High temperatures can
   A. denature an enzyme.
   B. increase an enzyme's molecular motion.
   C. change the protein structure of an enzyme.
   D. all of the above.

7. According to the induced-fit hypothesis
   A. the presence of the substrate causes the enzyme to adjust itself to the substrate, this creates stress on substrate bonds.
   B. enzymes and substrates fit perfectly together with "lock and key" precision.
   C. coenzymes alter the shape of enzyme molecules.
   D. inhibitors alter the shape of substrates.

8. Coenzymes
   A. are protein molecules.
   B. are not altered during a reaction.
   C. enable an enzyme to function.
   D. all of the above.

9. Which of the following is true?
   A. All enzymes work best at a neutral pH.
   B. There is an optimum pH for each specific enzyme.
   C. Enzymes work well at any pH higher than optimum.
   D. At a low pH, an enzyme needs more kinetic energy to function.

10. With negative-feedback inhibition, as the number of end products increases,
    A. enzyme activity decreases.
    B. enzyme activity increases.
    C. enzyme/substrate collisions become more effective.
    D. inhibitors attach to substrates.

11. Chemical messengers that tell the cell to decrease the production of a certain protein are
    A. coenzymes.
    B. gene-repressor proteins.
    C. inhibitors.
    D. denatured enzymes.

12. Enzymes function by
    A. lowering the activation energy required for a reaction.
    B. increasing the temperature of the reaction.
    C. providing activation energy to substrate molecules.
    D. increasing the production of substrate.

13. Inhibitor molecules
    A. react with end-products.
    B. attach to substrates.
    C. lower activation energy.
    D. attach to enzymes.

14. To increase the amount of end-product produced in an enzyme facilitated reaction, you could
    A. boil the enzyme.
    B. add inhibitor.
    C. add more substrate.
    D. add ice.

CHAPTER 6
BIOCHEMICAL PATHWAYS

Overview

Here we deal with some of the major chemical reactions that occur in living things. Because these reactions are dependent on one another and occur in specific series, they are commonly referred to as *biochemical pathways.* An understanding of these biochemical pathways will help you understand how energy is utilized within an organism. There are hundreds of such pathways, all of which interlink, but we will deal only with those that form the core of all chemical reactions in a living cell. The two major pathways are photosynthesis and cellular respiration.

Study Activities

1. Write a summary of each section of the *Chapter Outline* in your text.
2. For each of the *Learning Objectives* in your text, write a sentence or paragraph that demonstrates your mastery of the objective.
3. Answer the *Questions* at the end of the chapter in your text.
4. Complete the student study guide.

Key Terms/Notes

Define each of the following terms in the space provided, or make flash cards of the following terms.

Biochemical pathway

Light-energy conversion phase

Autotroph

Carbon dioxide conversion phase

Heterotroph

Photophosphorylation

Adenosine tri-phosphate

PGAL

High-energy phosphate bond

Aerobic cellular respiration

Chemiosmosis

Anaerobic

Glycolysis   Acetyl

Pyruvic acid   Electron transfer system

Krebs cycle   Protein respiration

Coenzyme A   Fermentation

Photosynthesis   FAD

Fat respiration   NAD$^+$

Questions with Short Answers

1. A plant uses energy from the _____ as a source of power for photosynthesis.

2. The waste product of photosynthesis is _____.

3. Raw materials for photosynthesis include _____ from the atmosphere and water.

4. Photosynthesis is associated with an organelle known as the _____.

5. Releasing energy from nutrients using oxygen is known as _____ cellular respiration.

6. The energy released from respiration is used to form high-energy chemical bonds in the molecule _____.

7. The first part of respiration, _____, is responsible for splitting a sugar into two parts.

8. The Krebs cycle takes place between the membranes of the _____.

9. One major event in the Krebs cycle is the removal of _____ and _____ from the sugars.

10. As one NADH progress completely through the electron transfer system, _____ ATP molecules can be formed.

11. The process of electron transfer in the ETS is controlled by _____ located on the inner membrane of the mitochondria.

12. A plant gets food by photosynthesis and then releases its energy by respiration, an animal gets its food by _____ and then releases its energy by respiration.

13. Oxygen is required for the _____ pathway of respiration.

14. Organisms are able to release energy from proteins, carbohydrates as well as _____.

15. The ultimate hydrogen ion and electron acceptor in fermentation is _____.

16. Yeast cells produce carbon dioxide and _____ as they engage in anaerobic cellular respiration.

17. The waste product of anaerobic cellular respiration of _____ cells is lactic acid.

Label/Diagram/Explain

Match L, C, G, K and E with the phrases below.

L = Light-energy conversion stage
C = Carbon dioxide conversion stage
G = Glycolysis
K = Krebs cycle
E = Electron transfer system

1. _____ produces O_2
2. _____ produces CO_2
3. _____ _____ ATP $\longrightarrow$ ADP + P
4. _____ _____ stages of photosynthesis
5. _____ produces pyruvic acid
6. _____ pyruvic acid is a raw material
7. _____ produces PGAL
8. _____ CO_2 is a raw material
9. _____ glucose is a raw material
10. _____ uses light energy

11. _____ H_2O is a raw material
12. _____ Uses O_2
13. _____ Produces a net of 2 ATP molecules from one glucose
14. _____ _____ _____ stages of cellular respiration
15. _____ occurs in cytoplasm
16. _____ occurs in grana
17. _____ occurs in stroma
18. _____ _____ occurs in mitochondria
19. _____ NADP + 2H $\longrightarrow$ $NADPH_2$
20. _____ _____ $NAD^+ + 2H$ $\longrightarrow$ NADH

21. Picture yourself as an atom of oxygen tied up in a molecule of carbon dioxide. Trace the pathway you would take if you were taken into a green plant that is undergoing photosynthesis. Then trace your pathways as you move into the process of aerobic cellular respiration. Be as specific as you can in describing your location and how you got there as well as the molecules of which you are a part.

Multiple Choice Questions

1. An autotroph is an organism that
   A. does not require food.
   B. does not respire.
   C. is able to "fix" carbon into an organic compound such as sugar.
   D. is a self starter.

2. Although ATP has three phosphate groups it has:
   A. only one bond.
   B. two high-energy bonds.
   C. enough places for four phosphates.
   D. the ability to hold an unlimited amount of energy.

3. The hydrogen which eventually becomes a part of sugar manufactured in photosynthesis is acquired from:
   A. water.
   B. active transport.
   C. sunlight.
   D. rearrangement of carbon dioxide molecules.

4. Light energy conversion provides the energy for the second phase of photosynthesis when:
   A. light is released as sound.
   B. water is cleaved into hydrogen and oxygen.
   C. sugar is used as the substrate.
   D. chemical bonds in ATP are formed.

5. The oxygen used for aerobic cellular respiration in plants comes from the process of
   A. photosynthesis.
   B. respiration.
   C. protein synthesis.
   D. chemical activity in mitochondria.

6. Fermenting yeast cells are able to release:
   A. as much energy from a sugar as humans.
   B. only two ATP (net) per glucose.
   C. carbon dioxide rather than oxygen.
   D. energy in the form of sugar substitutes.

7. When fats are respired two-carbon fragments from the fatty acids enter the:
   A. glycolysis pathway.
   B. photosynthetic pathway.
   C. Krebs cycle.
   D. electron transfer system.

8. Which of the following is the correct generalized chemical equation for photosynthesis?
   A. Energy + $C_6H_{12}O_6$ + $6O_2$ $\rightarrow$ $6CO_2$ + $6H_2O$
   B. Energy + $6CO_2$ + $6H_2O$ $\rightarrow$ $C_6H_{12}O_6$ + $6O_2$
   C. $C_6H_{12}O_6$ + $6O_2$ $\rightarrow$ $6CO_2$ + $6H_2O$ + Energy
   D. $6CO_2$ + $6H_2O$ $\rightarrow$ $C_6H_{12}O_6$ + $6O_2$ + Energy

9. Plants can use PGAL to
   A. make sugar molecules and lipids.
   B. make the carbon skeleton for amino acids.
   C. obtain energy.
   D. all of the above.

10. From one glucose, the entire aerobic cellular respiration pathway will net _____ molecules of ATP.
    A. 2
    B. 4
    C. 36
    D. 38

11. Which of these molecules is least like the others in terms of function?
    A. FAD
    B. PGAL
    C. NADP
    D. NAD^+

12. Cellular respiration occurs in
    A. autotrophs only.
    B. heterotrophs only.
    C. autotrophs and heterotrophs.
    D. animals but not plants.

13. The atmospheric oxygen released by plants comes from:
    A. H_2O.
    B. CO_2.
    C. PGAL.
    D. $C_6H_{12}O_6$.

14. ATP $\rightarrow$ ADP + P. This reaction represents
    A. energy stored.
    B. energy released.
    C. energy destroyed.
    D. energy created.

CHAPTER 7
DNA AND RNA: THE MOLECULAR BASIS OF HEREDITY

Overview

In previous chapters we considered a variety of biological structures and their functions. Organic molecules found in living cells are not haphazard arrangements of atoms, they are highly organized and can be classified into major groups. The group known as the *nucleic acids* has a unique structure and is the primary control molecule of the cell. The structure of these complex molecules is important in determining the actions of living cells.

Study Activities

1. Write a summary of each section of the *Chapter Outline* in your text.
2. For each of the *Learning Objectives* in your text, write a sentence or paragraph that demonstrates your mastery of the objective.
3. Answer the *Questions* at the end of the chapter in your text.
4. Complete the student study guide.

Key Terms/Notes

Define each of the following terms in the space provided, or make flash cards of the following terms.

Nucleic acid

Complementary bases

Nucleotide

Nucleoproteins

Nitrogenous base

Nucleosome/chromatin fibers

Ribose

Chromosome

Deoxyribose

DNA replication

Duplex DNA

Polymerase

Coding/Noncoding strands	Transfer RNA
Transcription	Ribosomal RNA
Central dogma	Translation
Anticodon	Transponson
Gene	Polysome
Codon	Chromosomal aberration
Mature RNA	Biotechnology
Messenger RNA	Recombinant DNA

Questions with Short Answers

1. The double helical structure is characteristic of _____ molecules.

2. The nitrogenous bases of DNA include: adenine, guanine, thymine, and _____ .

3. The nucleotides are attached to each other in a linear fashion by covalent bonds between the _____ and phosphate.

4. Complementary bases are attracted to each other in DNA by _____ bonds.

5. Cytosine hydrogen bonds to _____.

6. The _____ strand of DNA is the side upon which RNA may be formed.

7. Transcription results in the formation of _____ molecules.

8. Replication results in the formation of _____ molecules.

9. Translation results in _____ synthesis.

10. Each codon codes for one and only one _____.

11. Each tRNA _____ is complementary to a codon.

12. Transcription takes place in the _____ of the cell.

13. Changes in the sequence of nucleotide bases are called _____.

14. Gene manipulation or _____ in the future might control or cure genetic diseases.

15. The _____ is a lab procedure for copying selected segments of DNA.

16. Sections of mRNA called _____ are cut out so that mature mRNA can be produced.

17. _____ allow for the creation of new nucleotide sequences in existing DNA.

Label/Diagram/Explain

1. Create a base code for a single strand of DNA nine nucleotides long. Now construct the complementary strand. Did you use the base pairing rule? Use the first strand constructed to transcribe mRNA. Label the base sequence of the mRNA. Now show the anticodons of the tRNA which are complementary to the codons of the mRNA. Identify the amino acid each of the tRNAs will be attached to.

DNA: (complementary strand)	DNA: Gene	mRNA: Codons	tRNA: Anticodons	Amino acids specified
___	___	___	___	
___	___	___	___	___
___	___	___	___	
___	___	___	___	
___	___	___	___	___
___	___	___	___	
___	___	___	___	
___	___	___	___	___
___	___	___	___	

2. DNA & RNA - Basic Molecular Structure

Two main types of nucleic acids are deoxyribonucleic acid *a.*_____ and ribonucleic acid *b.*_____ . Each

of these are composed of subunits called *c.*_____ . A DNA nucleotide is composed

of a *d.*_____ group, *e.*_____ and one of four

nitrogenous bases: *f.*_____ , *g.*_____ , *h.*_____ or

*i.*_____ . An RNA nucleotide is composed of a *j.*_____ group,

*k.*_____ and one of four nitrogenous bases: *l.*_____ ,

*m.*_____ , *n.*_____ or *o.*_____ . Both DNA and

RNA molecules are polymers composed of long chains of nucleotides with the *p.*_____ of

one nucleotide bonding with the sugar of the next nucleotide in the chain. This forms the *q.*_____-

_____ backbone of the molecule. The DNA molecule is composed of two nucleotide strands. The

strands are held together by *r.*_____ bonds between bases. The bases

*s.*_____ and *t.*_____ attract one another and form two hydrogen bonds while

the bases *u.*_____ and *v.*_____ form three hydrogen bonds.

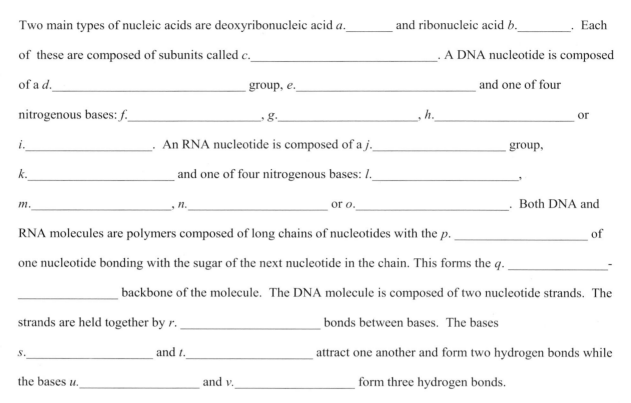

3. Translation
The assembly of amino acids into a polypeptide takes place at the *a.* _____ and is

called *b.* _____ . The first ribosomal subunit attaches to the *c.* _____ molecule which

carries the code for producing one protein. The second ribosomal subunit then moves into place. Cloverleaf

shaped *d.* _____ molecules carry *e.* _____ to the ribosome and *f.* _____ of the

tRNA molecule form temporary *g.* _____ bonds with *h.* _____ of the mRNA molecule.

A second tRNA molecule carries a second amino acid to the ribosome and hydrogen bonds again form between tRNA anticodons and mRNA codons. A *i.* _____ bond forms between the two amino acids that are brought into position by the tRNA molecules. This bond is the result of a

*j.*_____reaction which produces one water molecule. Next, a third tRNA molecule, carrying a third *k.* _____, moves into place and bonds with the *l.*

_____. As this happens, the first *m.* _____molecule releases from the mRNA. The amino acid it carried remains attached to the growing polypeptide chain. The amino acid carried by the third tRNA now forms a peptide bond with the second amino acid (now part of a dipeptide). A tripeptide and one

*n.*_____ molecule are produced.   The polypeptide continues to grow in a similar manner until the last amino acid is bonded into position. The protein molecule is then released from the ribosome.

4. Label the diagram below.

Multiple Choice Questions

1. One difference between preRNA and mature RNA is:
   A. kinds of nucleotide components.
   B. presence of extra RNA in preRNA.
   C. the part of DNA from which they were coded.
   D. only the age of the molecules.

2. Which of the following is NOT a chromosomal aberration?
   A. inversion
   B. duplication
   C. deletion
   D. X-ray

3. The site of protein synthesis is:
   A. at the nuclear membrane.
   B. at ribosomes .
   C. near microfilaments.
   D. always at a Golgi body.

4. While one strand of duplex DNA is being transcribed to mRNA:
   A. the complementary strand makes tRNA.
   B. the complementary strand is inactive.
   C. the complementary strand at this point is replicating.
   D. mutations are impossible during this short period.

5. One way to introduce new DNA into an organism is
   A. gene splicing.
   B. replication.
   C. removing introns.
   D. transcription.

6. Removing only one base in a DNA sequence:
   A. usually has no effect on the organism.
   B. could result in a chromosomal mutation.
   C. cannot occur without extremes of heat and pressure.
   D. can result in a significant change in the information about a protein.

7. A major difference between the genetic data of prokaryotic and eukaryotic cells is that in prokaryotes the:
   A. genes are RNA not DNA.
   B. histones are arranged differently.
   C. duplex DNA is circular.
   D. duplex DNA is absent in bacteria.

8. If the DNA gene strand has the base sequence CCA - TAT - TCG, the complementary DNA strand will have the sequence:
   A. CCA - TAT – TCG.
   B. GGU - AUA – AGC.
   C. CCA - UAU – UCG.
   D. GGT - ATA – AGC.

9. A DNA gene strand with the base sequence CCA — TAT — TCG will be transcribed into RNA with the base sequence:
   A. CCA - TAT – TCG.
   B. GGU - AUA – AGC.
   C. CCA - UAU – UCG.
   D. GGT - ATA – AGC.

10. A DNA gene strand with the base sequence CCA — TAT — TCG codes for the amino acid sequence: (consult the Amino Acid — Nucleic Acid Dictionary in your text or lab manual)
    A. proline - tyrosine – serine.
    B. glycine - isoleucine – threonine.
    C. proline- tyrosine – threonine.
    D. glycine - isoleucine – serine.

11. The mRNA codon CAU will form temporary bonds with the
    A. mRNA anticodon CAU.
    B. mRNA codon GUA.
    C. tRNA anticodon GUA.
    D. tRNA codon CAU.

12. If the DNA base sequence GAG is mutated to GAC (consult the Amino Acid — Nucleic Acid Dictionary in your text)
    A. aspartic acid will substitute for glutamic acid in the resulting polypeptide.
    B. the resulting protein will be unable to function.
    C. there will be no change in the amino acid sequence of the resulting polypeptide.
    D. a chromosomal mutation has occurred.

13. In eukaryotic cells, mature RNA is formed by the
    A. removal of introns.
    B. removal of exons.
    C. addition of introns.
    D. addition of exons.

CHAPTER 8
MITOSIS: THE CELL-COPYING PROCESS

Overview

You have seen how the molecule DNA replicates. Once this process is complete, replicated DNA is distributed to two new daughter cells by the process of mitosis. The way this cell-splitting process occurs assures that the daughter cells will have the same genetic message as the original DNA. Mitosis is a sequence of events that is separated into stages in order to better visualized these complex events. However, even though the resulting daughter cells have identical genetic messages, they may differ in the way they are built and the specific roles they perform. The relationships among DNA, genes, and chromosomes are concepts for later consideration of genetics, evolution, and sex-cell formation.

Study Activities

1. Write a summary of each section of the *Chapter Outline* in your text.
2. For each of the *Learning Objectives* in your text, write a sentence or paragraph that demonstrates your mastery of the objective.
3. Answer the *Questions* at the end of the chapter in your text.
4. Complete the student study guide.

Key Terms/Notes

Define each of the following terms in the space provided, or make flash cards of the following terms.

Mitosis   Centromere

Cytokinesis   Centrioles

Interphase   Spindle

Chromosomes   Metaphase

Prophase   Anaphase

Chromatid   Daughter chromosomes

Telophase   Cell plate

Daughter nuclei   Differentiation

Daughter cells   Cancer

Cleavage furrow   Malignant

Apoptosis   Metastasis

Questions with Short Answers

1. The chromosomes align at the cell's equator during the _____ stage of mitosis.

2. In interphase the original duplex DNA replicates and two identical DNA molecules known as _____ are formed.

3. During anaphase the _____ aids in the movement of the chromosomes to the cell's poles.

4. Animal cells divide into two cells as the result of an indentation known as the _____ furrow.

5. _____ is the the process that occurs when a cell adopts special characteristics.

6. If the mechanism that controls cell division malfunctions, cell division continues and may result in _____.

7. Cell division results in the formation of two identical _____ cells.

8. During anaphase the _____ splits and the chromosomes move to the poles.

9. When a cell is in the prophase stage of mitosis, the _____ membrane disintegrates.

10. Mitosis is basically the same in plant and animal cells. The only major difference between the two is how they undergo the process of _____.

11. When tumors have cells break off of the original mass and form new tumors in other regions of the body, _____ has occurred.

12. _____ is the stage in the cell cycle when growth occurs.

13. _____ normally occurs in many cells of the body because they might be harmful or it takes too much energy to maintain them.

Label/Diagram/Explain

Draw and label the four stages of mitosis for an animal cell with three pairs of chromosomes. Show the arrangement of the chromosomes in each stage, identify cellular structures and list the major events that occur at each stage. Your drawings would include the following where appropriate:

chromosomes	spindle
sister chromatids	nuclear membrane
daughter chromosomes	nucleolus
centromeres	cleavage furrow
centrioles	

Stage: _____

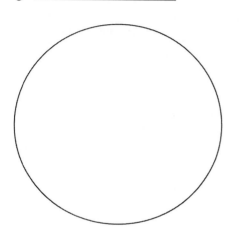

Major events:

Stage: _____

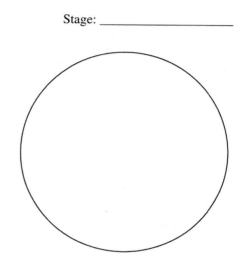

Major events:

Stage: _____

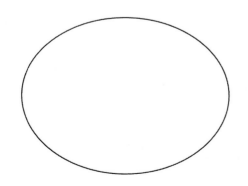

Major events:

Stage: _____

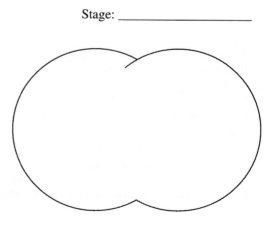

Major events:

Multiple Choice Questions

1. The centromeres split during
   A. anaphase.
   B. prophase.
   C. metaphase.
   D. interphase.

2. What would happen if microtubules where prevented from forming during mitosis?
   A. the cell plate would not form
   B. replication would not occur
   C. centromeres would not split
   D. anaphase cannot occur

3. The normal state of chromosomes in prophase is as
   A. daughter chromosomes.
   B. chromosomes composed of two chromatids.
   C. chromatids composed of two chromosomes.
   D. chromosomes consist of single chromatids.

4. The presence of cell walls in plants is associated with _____ in telophase.
   A. cleavage furrows
   B. spindle fiber formation
   C. differentiation
   D. cell plate formation

5. During which stage of the cell cycle does DNA replication occur?
   A. the S stage of interphase
   B. anaphase of mitosis
   C. G_2 stage of metaphase
   D. prophase

6. Nerve cells do not normally undergo mitosis. This means that
   A. the brain is unimportant.
   B. your brain cannot grow.
   C. cytokinesis will be common in nerve tissue.
   D. transcription of DNA will not occur.

7. Radiation is able to successfully control cancer because
   A. cancer cells do not grow rapidly.
   B. cancer cells spend most of their time in the S stage of prophase.
   C. it stimulates programmed cell death.
   D. these agents only affect diseased cells.

8. The correct order for the stages of mitosis is
   A. prophase - anaphase - metaphase – telophase.
   B. metaphase - prophase - anaphase – telophase.
   C. prophase - metaphase - anaphase – telophase.
   D. prophase - metaphase - telophase – anaphase.

9. Compared to the mother cell, daughter cells at the end of mitosis have _____ number of chromosomes.
   A. half the
   B. the same
   C. twice the
   D. none of the above

10. Which one of the following is not an event of telophase?
    A. nucleoli reappear
    B. spindle disappears
    C. daughter nuclei form
    D. centrioles duplicate

11. During anaphase
    A. chromosomes become visible.
    B. daughter chromosomes migrate to the poles.
    C. chromosomes line up at the equatorial plane.
    D. cytokinesis is completed.

12. Cell division is needed for
    A. growth.
    B. replacement of warn-out cells.
    C. healing of damaged tissue.
    D. all of these.

13. Chromosomes are composed of sister chromatids during
    A. interphase.
    B. prophase only.
    C. prophase and metaphase.
    D. prophase, metaphase, and anaphase.

14. Sister chromatids contain
    A. identical strands of DNA.
    B. half of a duplex DNA molecule.
    C. different genetic information.
    D. one gene.

CHAPTER 9
MEIOSIS: SEX-CELL FORMATION

Overview

How can the chromosome number in humans remain at forty-six generation after generation if both parents contribute equally to the genetic information of the child. The answer is found in the process of meiosis. Meiosis is a specialized cell division that results in the formation of sex cells. Knowing the mechanics of this process is essential to understanding how genetic variety can occur in sex cells. This variety ultimately shows up as differences in offspring.

Study Activities

1. Write a summary of each section of the *Chapter Outline* in your text.
2. For each of the *Learning Objectives* in your text, write a sentence or paragraph that demonstrates your mastery of the objective.
3. Answer the *Questions* at the end of the chapter in your text.
4. Complete the student study guide.

Key Terms/Notes

Define each of the following terms in the space provided, or make flash cards of the following terms.

Sexual reproduction

Haploid

Sperm

Diploid

Eggs

Crossing-over

Gamete

Homologous chromosomes

Gametogenesis

Gonads

Zygote

Ovaries

Testes   Independent assortment

Pistil   Autosomes

Anther   Chromosomal abnormalities

                                               Nondisjunction
Reduction division

                                               Trisomy/Monosomy
Synapsis

                                               Down's syndrome
Segregation

Sex chromosomes

Questions with Short Answers

1. _____is the switching of equal portions of DNA between two homologous chromosomes.

2. During meiosis in human cells, each of the twenty-three pairs of chromosomes undergoes segregation. The fact that each pair segregates without any regard to the other pairs in known as _____ .

3. In meiosis the haploid number of chromosomes first appears in _____.

4. Daughter cells formed by _____have the same genetic composition as the original cell.

5. In humans, if a Y bearing sperm fertilizes an X bearing egg, the offspring will be a _____.

6. Fertilization results in the formation of a cell with the _____ number of chromosomes.

7. In males, meiosis occurs in organs called the _____.

8. The _____ meiotic division results in the formation of daughter cells with the haploid number of chromosomes.

9. In mitosis and meiosis cytokinesis occurs in the _____ stages.

10. Synapse occurs in the _____ stage.

11. Crossing-over occurs during the _____ of meiosis.

12. Crossing-over, fertilization, and independent assortment all result in increased genetic _____.

13. Daughter chromosomes are formed in the _____ stage of meiosis.

14. _____ are organs where meiosis may occur.

15. An individual displaying Down's syndrome has a condition known as _____.

16. _____ occurs if a pair of homologous chromosomes does not segregate properly during gametogeneis.

Label/Diagram/Explain

1. In mitosis and meiosis there are events that are common to both processes. List events that occur in both cell division processes. Include each stage in your answer.

2. There are also differences between the two processes. Describe these differences.

3. Diagram and label the stages of meiosis for a cell with two pairs of chromosomes. Pay special attention to the number of chromosomes present at each stage. Also note whether the chromosomes are composed of two duplex DNA molecules (sister chromatids) or composed of a single duplex DNA molecule.

MEIOSIS I   **MEIOSIS II**

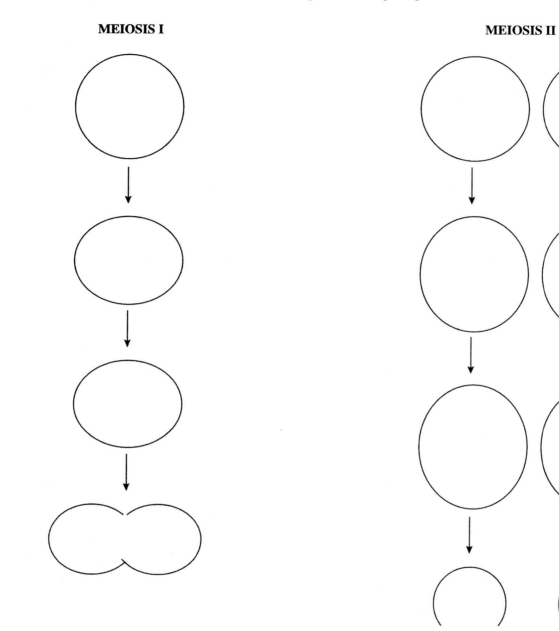

Multiple Choice Questions

1. The exchange of genetic material (genes) between segments of homologous chromosomes results in
   A. new gene combinations.
   B. zygotes.
   C. diploid cells.
   D. segregation of genes.

2. A process that occurs during prophase I is
   A. segregation.
   B. synapsis.
   C. reduction division .
   D. independent assortment.

3. The diploid number of chromosomes is found in cells during:
   A. prophase II.
   B. telophase I .
   C. anaphase II.
   D. prophase I.

4. The fact that each homologous pair of chromosomes in humans separates and moves to the poles without being influenced by the other pairs is
   A. segregation.
   B. disintegration.
   C. independent assortment.
   D. fertilization.

5. A new nuclear membrane is formed in:
   A. anaphase I.
   B. prophase II.
   C. telophase I.
   D. anaphase II.

6. In mitosis the centromeres split during anaphase; and, in meiosis during:
   A. anaphase I.
   B. telophase I .
   C. prophase II.
   D. anaphase II .

7. Diploid cells are formed by:
   A. synapsis.
   B. reduction division.
   C. fertilization .
   D. independent assortment.

8. An organism having a diploid number of 12 forms gametes having:
   A. 6 chromosomes.
   B. 12 chromosomes .
   C. 18 chromosomes .
   D. 24 chromosomes.

9. Segregation of homologous chromosomes occurs during
   A. mitosis.
   B. meiosis I.
   C. meiosis II.
   D. all of the above.

10. A gamete can have too many or too few chromosomes when _____ occurs.
    A. nondisjunction.
    B. crossing over.
    C. synapsis.
    D. independent assortment.

11. Variation among offspring in sexually reproducing organisms comes about by
    A. crossing over.
    B. fertilization.
    C. independent assortment.
    D. all of the above.

12. Gametogenesis produces
    A. sex cells.
    B. gonads.
    C. zygotes.
    D. testes.

13. Crossing over occurs between
    A. sister chromatids.
    B. homologous chromosomes.
    C. gametes.
    D. spindle fibers.

14. Which of the following represents chromosome number before and after the process of meiosis?
    A. n $\rightarrow$ n
    B. n $\rightarrow$ 2n
    C. 2n $\rightarrow$ n
    D. 2n $\rightarrow$ 2n

CHAPTER 10
MENDELIAN GENETICS

Overview
You have been introduced to the concepts and importance of DNA as a molecule for storing the genetic information used to manufacture proteins and to guide the processes of mitosis and meiosis. Gregor Mendel first established the principles associated with the transmission of characteristics from one generation to another. His work was the foundation of current genetic thinking. Since his discoveries, DNA has been identified as the genetic material that is passed from parent to offspring. This knowledge allows scientists to predict the results of breeding and to develop new methods of controlling the genetic makeup of many types of organisms.

Study Activities
1. Write a summary of each section of the *Chapter Outline* in your text.
2. For each of the *Learning Objectives* in your text, write a sentence or paragraph that demonstrates your mastery of the objective.
3. Answer the *Questions* at the end of the chapter in your text.
4. Complete the student study guide.

Key Terms/Notes
Define each of the following terms in the space provided, or make flash cards of the following terms.

Gene

Genotype

Genetics

Phenotype

Mendelian genetics

Homozygous

Alleles

Heterozygous

Locus

Dominant allele

Genome

Recessive allele

Mendel's law of dominance

Multiple alleles

Mendel's law of segregation

Polygenic inheritance

Mendel's law of independent assortment

Pleiotropy

Probability

Linkage group

Single-factor cross

Autosomes

Punnett square

Sex chromosomes

Double-factor cross

X - linked

Codominance

Expression

Questions with Short Answers

1. The result of gene expression is called the _____ of the organism.

2. A person having two alleles for type A blood would be _____ for that trait.

3. When some red flowers are crossed with white flowers the offspring are always pink. In these plants the condition for flower color shows a _____.

4. If you roll four dice, the probability of rolling all fives is _____.

5. That portion of a DNA strand that contains the information for a particular trait is called a(n) _____.

6. If both parents express a dominant characteristic and one of their offspring shows the recessive condition, both parents are _____ for the trait.

7. When four separate pairs of genes, located on four different chromosomes, determine a single trait, the trait is a result of _____ .

8. Alleles that are inheriteted on the X chromosomes are _____ .

9. The _____ is the specific region on the DNA were a particular gene is found.

10. The fact that some cat's ears develop dark pigment because they have a temperature that is lower than the rest of their body is an example of the influence of _____ on the genes.

11. If one parent exhibits a recessive trait and the other the dominant, and all the offspring show the recessive trait, the second parent is _____ for the dominant condition.

12. A genetics problem dealing with a person's hairline (straight or widow's peak) and earlobe shape (free or attached) is a _____ cross.

13. The _____ gene is found on the Y chromosome.

14. Genes found on the X-Chromosome are called _____.

15. A _____ is any person who is heterozygous for a trait; the recessive allele is hidden, and does not express itself enough to be a phenotype.

Label/Diagram/Explain

At a large family reunion people were mentioning the fact that everyone present had curly black hair and dark brown eyes. Several hours later a long forgotten cousin, Fred, arrived with his wife Sarah and their ten children.

Great grandmother could not understand why Fred had straight red hair and green eyes; no one in the family ever had these traits. She also wanted to know why Sarah and (1) five of the children had curly black hair and brown eyes; (2) three children had curly red hair and green eyes; (3) two of the children had Fred's straight red hair and brown eyes.

You are a college-educated member of the family. Explain to grandmother how this could occur.

Multiple Choice Questions

1. BbCCDd is the genotype of an organism. How many different types of gametes can this organism produce?
   A. 1
   B. 2
   C. 3
   D. 4

2. An example of a phenotype is
   A. AB type blood.
   B. allele for type A and B blood.
   C. a sperm with an allele for A blood.
   D. lack of iron in the diet causing anemia.

3. Body size is determined by the interaction of numerous alleles. This is an example of
   A. autosomes.
   B. pleiotropy.
   C. single-factor crosses.
   D. polygenic inheritance.

4. The probability of guessing the correct answer to this question is
   A. four .
   B. one fourth.
   C. sixteen.
   D. 1/4 times 1/4.

5. If both parents are heterozygous, the probability that the recessive trait will appear in the offspring is
   A. 1/2.
   B. 1/4.
   C. 3/4.
   D. zero.

6. In order for a recessive X-linked trait to appear in a female, she must inherit a recessive allele from
   A. neither parent.
   B. both parents .
   C. her father only.
   D. her mother only.

7. It was noticed that in certain flowers, if a flower had red petals they were usually small petals. If the petals were white, pink, or orange they were usually large petals. This could be the result of
   A. lack of dominance.
   B. pleiotropy.
   C. linkage groups.
   D. polygenic inheritance.

8. In pea plants, tall plant height (T) dominates dwarf (t). If a plant heterozygous for the height trait is crossed with a dwarf plant, what will be the phenotypic ratios of the offspring?
   A. all Tt
   B. 1/2 Tt, 1/4TT, 1/4tt
   C. 1/2 tall, 1/2 dwarf
   D. 3/4 tall, 1/4 dwarf

9. A man with normal color vision marries a woman with normal vision, but who is a carrier for color blindness. The probability that their first child will be color blind is _____ ; the probability that their first daughter will be color blind is _____; and the probability that their first son will be color blind is _____.
   A. 1/2; 1/2; 1/2
   B. 1/4; 0; 1/2
   C. 1/2; 0; 1/2
   D. 1/4; 1/2; 1/2

10. A woman with blood type O and a man with blood type AB have a child together. What are the possible bloodtypes of this child?
    A. AB or O
    B. A, B, or O
    C. A or B
    D. AB, A, B, or O

11. A radish plant that produces round radishes is crossed with a radish plant that produces long radishes. All of the offspring have oval radishes. Radish shape may be inherited by
    A. polygenic inheritance.
    B. pleiotropy.
    C. multiple alleles.
    D. lack of dominance.

12. In the following cross, what is the probability that the offspring will exhibit both dominant traits?
    Aabb × aaBB

    A. 1/2
    B. 3/8
    C. 4/16
    D. 1

13. "She has really long fingers and toes and is exceptionally tall." This is a statement of
    A. genotype.
    B. phenotype.
    C. monohybridization.
    D. locus placement.

49

CHAPTER 11
DIVERSITY WITHIN SPECIES

Overview

In your travels you may have noticed that plants and animals of the same kind may vary slightly in different regions; and that we artificially maintain certain groups of characteristics in domesticated species. It is important to understand how genetic variety is introduced and maintained within a population.

Study Activities

1. Write a summary of each section of the *Chapter Outline* in your text.
2. For each of the *Learning Objectives* in your text, write a sentence or paragraph that demonstrates your mastery of the objective.
3. Answer the *Questions* at the end of the chapter in your text.
4. Complete the student study guide.

Key Terms/Notes

Define each of the following terms in the space provided, or make flash cards of the following terms.

Species

Strains

Population

Varieties

Gene frequency

Clones

Gene pool

Hybrid

Demes

Monoculture

Subspecies

Genetic counselor

Races

Eugenic laws

Breeds

Questions with Short Answers

1. Animals that can naturally reproduce fertile offspring belong to the same _____.

2. A pond has a large population of black bass. All of the genes found in these bass belong to the same _____.

3. If you take a leaf from a plant and use it to start another plant, the two plants are _____.

4. Laws that are passed in an attempt to improve the human race by controlling who may marry and who may have children are _____.

5. When two different strains of a plant are crossed, the offspring are known as _____.

6. A _____ is a local population that has gene frequencies somewhat different from other local populations of the species.

7. A single type of cotton is being grown in large fields. This is an example of _____ .

8. You would probably visit a _____ if you had a question concerning an inherited disorder in your family.

9. All of the cats living within a city are referred to as a _____ of cats.

10. _____ are a means of getting new genes into a population.

11. _____ reproduction provides new genotypes in the offspring.

12. Collies and beagles are two breeds of dogs. However, the individual collies and beagles all belong to the same _____.

13. Clones are individuals having the same _____.

14. Within a _____, genes are repackaged into new individuals from one generation to the next.

15. A term that is sometimes used to describe the degree of genetic difference among individuals within a population is _____.

16. The _____ concept is an attempt to define groups of organisms that are reproductively isolated and, therefore, constitute a distinct unit of evolution.

Label/Diagram/Explain

You are a breeder of a particular type of dog. In one of the litters you discover a male with a unique coat which has never before been seen in this breed. When it becomes known that you have a dog with this color, everyone wants to buy one of these animals at triple the usual price.

How would you go about producing as many dogs as possible with this color? What would you do to make certain that no one else would be able to breed the dogs you sold?

Multiple Choice Questions

1. Which of the following would demonstrate the greatest genetic variety?
   A. demes
   B. hybrids
   C. a species
   D. clones

2. If every time two animals are crossed they produce offspring that can not reproduce, the two animals
   A. are members of the same species.
   B. belong to two different subspecies.
   C. are members of two different species.
   D. are both sterile.

3. If you wish to produce a large number of plants that have the same appearance, you should reproduce them
   A. by making clones.
   B. sexually.
   C. as hybrids.
   D. from seeds.

4. The migration of animals from one area to another may
   A. result in new species.
   B. result in extinction.
   C. change the local gene pools.
   D. result in fewer recessive alleles.

5. Eliminating a specific human genetic disease is difficult if not impossible because
   A. most diseases are caused by dominant alleles.
   B. mutation rates are high.
   C. immigration introduces new alleles.
   D. controlling reproductive behavior is difficult.

6. Genetic variety can be added to a population by
   A. mutations.
   B. migration.
   C. sexual reproduction.
   D. all of the above.

7. As a bird flies over a small remote island it defecates and drops seeds which later germinate. Plants from these seeds mature and reproduce sexually. After several generations 2000 decedents of these original plants live on the island. These decedents represent
   A. a new species.
   B. clones.
   C. hybrids.
   D. a deme.

8. Which of the following is true concerning recessive alleles?
   A. recessive alleles are always present in a population at a lower frequency than dominant alleles
   B. uncommon traits are the expression of recessive alleles and common traits are the expression of dominant alleles
   C. beneficial traits are the expression of dominant alleles and deleterious traits are the expression of recessive alleles
   D. Recessive traits can be common or uncommon; beneficial or deleterious.

9. For a group of animals to belong to the same species they must
   A. look alike.
   B. live in the same geographic region.
   C. have the ability to produce fertile offspring.
   D. all of the above.

10. "In one human population the allele for blood type O comprises 95% of all alleles for blood type." This statement expresses the concept of
    A. allele frequency.
    B. gene pool.
    C. deme.
    D. Eugenics.

11. Eugenics laws were unsuccessful because
    A. the "human condition" is influenced by the environment as well as genes.
    B. many genetic abnormalities are caused by recessive genes which are carried and "hidden" in heterozygous individuals.
    C. most genetic conditions are not inherited in a simple dominant/recessive fashion.
    D. all of the above.

12. Variety is to plant as _____ is to bacteria.
    A. species
    B. strain
    C. race
    D. clone

13. Which one of the following encourages the maintenance of demes?
    A. migration
    B. geographic barriers
    C. mutations
    D. hybrid crosses

CHAPTER 12
NATURAL SELECTION AND EVOLUTION

Overview

Previously background was presented in the areas of chemistry, information systems (DNA), sexual reproduction, heredity, and population genetics. These are all closely related to one another and are also related to the surroundings of organisms. Since the surroundings are always changing, the survival of living things depends on their ability to adjust their processes to these changes. The changes that assure survival can occur to any individual in the population, but unless they are genetically (DNA) determined and transmitted to the next generation, they will be of little value to the survival of the species. It is important to identify how differences come about and how they may change a sexually reproducing species over thousands of generations.

Study Activities

1. Write a summary of each section of the *Chapter Outline* in your text.
2. For each of the *Learning Objectives* in your text, write a sentence or paragraph that demonstrates your mastery of the objective.
3. Answer the *Questions* at the end of the chapter in your text.
4. Complete the student study guide.

Key Terms/Notes

Define each of the following terms in the space provided, or make flash cards of the following terms.

Genetic recombination

Natural selection

Genome

Differential reproduction

Spontaneous mutations

Hardy-Weinberg law

Acquired characteristic

Selecting agent

Theory of natural selection

Evolution

Questions with Short Answers

1. Natural changes in the DNA of a cell are called _____.

2. A complete set of genes for an organism is known as a _____.

3. Characteristics displayed by organisms that are not the result of gene action are called _____ characteristics.

4. Every individual that is produced by sexual reproduction is unique because of genetic _____.

5. From the point of view of natural selection, when a predator kills and eats a prey organism the predator is acting as a _____.

6. The _____ states that gene frequencies will remain unchanged if random mating occurs, no mutation occurs, a large population is present, and no migration occurs.

7. Those individuals that have larger numbers of offspring will pass more of their genes on to the next generation. This is an example of _____.

8. If POPULATION A has many individuals with curly hair and POPULATION B has none, POPULATION A has a _____ frequency of the curly hair allele.

9. Most populations of most species have a natural capacity to produce _____ offspring than what is needed to replace the parents when they die.

10. New genes can enter a population through _____ and _____.

11. The genetic adaptation of a population to its environment is _____.

12. Gene-frequency differences that result from chance are more likely to occur in _____ populations than in _____ populations.

13. Maintaining _____ diversity in the population can be very difficult when the species consists of few individuals.

14. If one individual leaves 100 offspring and another leaves only 2, the first organism has passed more copies of its _____ to the next generation than has the second.

Label/Diagram/Explain

You are a broccoli farmer and you have a problem with aphids, which are insects that suck juices from your broccoli plants. Aphids are insects that have a very large reproductive capacity. However, most of their reproduction is asexual with females laying unfertilized diploid eggs that have the same exact gene combination as the mother. Ladybird beetles enjoy eating aphids, so you introduce some of these predators into some of your broccoli fields. In the other broccoli fields you use an insecticide you have used for the past 10 years to control the aphids and find that it does not work as well as it has in the past. As you walk about your fields you notice that some plants are heavily infested while others are not. When you transfer aphids from heavily infested plants to those plants that have no aphids, the plants to which you transferred aphids do not become infested.

Based on what your have learned about how natural selection works and how gene frequencies can be altered, answer the following questions:

1. What are two selecting agents acting on the aphid population?

2. Which of the kinds of organisms discussed is likely to have the lowest genetic variety? why?

3. Why doesn't the insecticide work as well as it once did?

4. What is a possible reason for the fact that some broccoli plants do not become infested with aphids?

Multiple Choice Questions

1. Which one of the following would result in a new allele being introduced into a population?
   A. a mutation in a skin cell
   B. the addition of new individuals to the population by reproduction
   C. migration out of the population
   D. a mutation in the gamete producing cells

2. Which one of the following would NOT allow an allele to be hidden?
   A. an allele might be recessive
   B. an individual might die before the allele had an opportunity to act
   C. the allele could be an acquired characteristic
   D. the allele might only express itself under certain environmental conditions

3. Which one of the following is a common method by which natural selection results in altered gene pools?
   A. some individuals have genes that allow them to have larger numbers of offspring
   B. some individuals are accidentally killed
   C. some individuals do not carry mutations
   D. some individuals are members of very large populations

4. In order to keep allele frequencies constant it is necessary to
   A. have small populations.
   B. have totally random mating.
   C. allow for the free movement of individuals in and out of the population.
   D. increase the mutation rate.

5. If the frequency of the blue eye allele in a population in 1800 was 50% and the frequency in 1990 was 47%, which one of the following could be correct?
   A. the frequency of the blue eye allele increased
   B. blue-eyed individuals may have had larger numbers of offspring
   C. blue-eyed individuals may have migrated from the population
   D. blue-eyed individuals had no mutations

6. Evolution is the result of
   A. random mating.
   B. natural selection.
   C. the inheritance of acquired characteristics.
   D. the existence of Hardy-Weinberg conditions.

7. Air pollution from the Industrial Revolution darkened the bark of trees in England. For the peppered moth, this change in the environment resulted in
   A. differential survival.
   B. a change in allele frequencies.
   C. an increase in the occurrence of dark moths.
   D. all of the above.

8. An acquired characteristic would be
   A. a woman's hair that turns gray at age 45.
   B. the rapid height increase of a teenage boy.
   C. a child learning to speak.
   D. dimples.

9. The Theory of Natural Selection was proposed
   A. independently by Charles Darwin.
   B. jointly by Darwin and Wallace.
   C. independently by Gregor Mendel.
   D. jointly by Hardy and Weinberg.

10. In a certain population, 80% of the alleles for a given gene are dominant and 20% of the alleles are recessive. What percentage of the population would you expect to be heterozygous if Hardy-Weinberg conditions exist?
    A. 16%
    B. 25%
    C. 32%
    D. 50%

11. Gene frequencies can change when
    A. individuals with certain phenotypes have a greater survival rate than individuals with other phenotypes.
    B. individuals with certain phenotypes produce more offspring than individuals with other phenotypes.
    C. individuals with certain phenotypes are chosen as mates more frequently than individuals with other phenotypes.
    D. all of the above.

12. Which of the following is not a Hardy-Weinberg condition?
    A. no migration
    B. no mutations
    C. random mating
    D. small population size

CHAPTER 13
SPECIATION AND EVOLUTIONARY CHANGE

Overview

Most scientists accept the theory that plant and animal species have changed from their first appearance on earth and continue to change today. Some important questions can be raised about this theory. (1) How do species change? (2) What causes new species to be formed? (3) What evidence exists that new species have been produced? There are several kinds of evidence that support the concept of speciation.

Study Activities

1. Write a summary of each section of the *Chapter Outline* in your text.
2. For each of the *Learning Objectives* in your text, write a sentence or paragraph that demonstrates your mastery of the objective.
3. Answer the *Questions* at the end of the chapter in your text.
4. Complete the student study guide.

Key Terms/Notes

Define each of the following terms in the space provided, or make flash cards of the following terms.

Gene flow

Reproductive isolating mechanisms

Range

Genetic isolating mechanisms

Species

Habitat preference

Geographic isolation

Ecological isolation

Geographic barriers

Seasonal isolation

Subspecies

Behavioral isolation

Speciation

Polyploidy

Divergent evolution   Gradualism

Adaptive radiation   Punctuated equilibrium

Convergent evolution

Questions with Short Answers

1. The concept that evolutionary change can proceed rapidly for a time followed by periods of little change is called _____.

2. A situation in which distinctly separate species come to resemble each other is known as _____.

3. _____ results in the rapid evolution of a large number of closely related species.

4. _____ is the earliest humanlike organism known from the fossil record.

5. In most cases the process of speciation probably requires that _____ separate two or more portions of a species from one another.

6. A _____ is a population within a species that is distinct but still capable of interbreeding with other similar organisms within the species.

7. A group of organisms that can interbreed naturally and produce fertile offspring is a _____.

8. Habitat preference, seasonal isolation, and behavioral isolation are examples of _____ mechanisms.

9. The distribution of genes to new geographic regions and to new generations is known as _____.

10. _____ is a condition of having multiple sets of chromosomes rather than the normal haploid or diploid number.

11. Traditionally biologists assumed that the process of evolution took place slowly as changes accumulated. This concept is known as _____.

12. Some people consider _____ to be a race of humans specially adapted to life in the harsh conditions found in post-glacial Europe. Others consider them to be a separate species.

13. About 800,000 years ago, archaic forms identified as _____ were identified in the fossil record.

Label/Diagram/Explain

Describe what each of the four illustrations below represents. Explain how you would know that the new groups in the fourth illustration are really different species and not just geographic variations or subspecies.

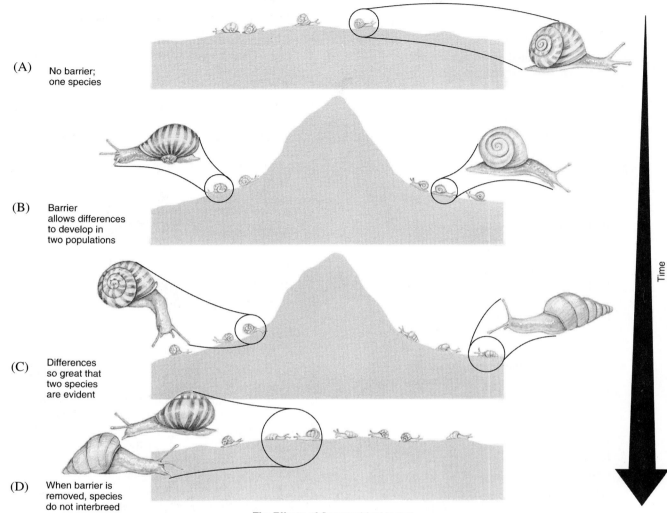

(A)   No barrier;
      one species

(B)   Barrier
      allows differences
      to develop in
      two populations

(C)   Differences
      so great that
      two species
      are evident

(D)   When barrier is
      removed, species
      do not interbreed

Time

The Effects of Geographical Isolation

Multiple Choice Questions

1. Hybrid animals like mules are not considered to be a species because
   A. they do not reproduce.
   B. they are not common enough.
   C. they can only be maintained by humans.
   D. they are the result of convergent evolution.

2. Two closely related organisms are not considered to be separate species unless
   A. they look different.
   B. they are reproductively isolated.
   C. they are able to interbreed.
   D. they are in different geographic parts of the world.

3. The Darwin-Wallace theory of natural selection differs from Lamarck's ideas in that
   A. Lamarck understood the role of genes and Darwin and Wallace did not.
   B. Lamarck assumed that characteristics obtained during an organism's lifetime could be passed to the next generation, Darwin and Wallace did not.
   C. Lamarck did not think that evolution took place, Darwin and Wallace did.
   D. Lamarck developed the basic ideas of speciation which Darwin and Wallace refined.

4. What percent of the number of species that have ever been present on the face of the earth are extinct?
   A. 10 percent.
   B. 50 percent.
   C. 75 percent.
   D. 99 percent.

5. Which of the following is NOT necessary for speciation?
   A. genetic isolation from other species
   B. genetic variety within a species
   C. hundreds of millions of years
   D. reproduction

6. Hummingbirds and honeybees both make noise as they fly from flower to flower collecting nectar. This is an example of
   A. adaptive radiation.
   B. convergent evolution.
   C. divergent evolution.
   D. punctuated equilibrium.

7. Polyploidy
   A. is common among animals.
   B. is generally a slow evolutionary process.
   C. involves multiple sets of chromosomes.
   D. is a rare mechanism for speciation.

8. Punctuated equilibrium was proposed by
   A. Darwin and Wallace.
   B. Lamarck.
   C. Eldredge and Gould.
   D. Buffon.

9. An organism invades an unexploited environment. What type of evolutionary pattern would you expect will follow?
   A. divergent evolution
   B. adaptive radiation
   C. polyploidy
   D. convergent evolution

10. Although several species of firefly may live in the same area, they do not interbreed. Each species of firefly has a unique flash that is used to signal potential mates of the same species. In fireflies, cross-species matings are prevented by
    A. mechanical isolation.
    B. ecological isolation.
    C. seasonal isolation.
    D. behavioral isolation.

11. Two groups of organisms belong to different species if
    A. gene flow between the two groups is not possible, even in the absence of physical barriers.
    B. physical barriers separate the two groups thereby preventing cross matings.
    C. the two groups of organisms have a different physical appearance.
    D. individuals from the two groups, when mated, produce hybrid offspring.

12. The Darwin-Wallace theory of natural selection corresponds with
    A. gradualism.
    B. polyploidy.
    C. punctuated equilibrium.
    D. the inheritance of acquired characteristics.

CHAPTER 14
ECOSYSTEM ORGANIZATION AND ENERGY FLOW

Overview

All living things require a continuous source of energy, which they use for growth, movement, reproduction, and many other activities. Certain physical laws describe how energy changes occur. The second law of thermodynamics states that during the process of converting energy from one form to another, some useful energy is lost as useless heat. Many of the world's problems result from our failure to recognize the limits imposed by the laws of thermodynamics. That energy is used and converted within groups of interacting organisms, and the laws of thermodynamics apply to living systems is a important concept.

Study Activities

1. Write a summary of each section of the *Chapter Outline* in your text.
2. For each of the *Learning Objectives* in your text, write a sentence or paragraph that demonstrates your mastery of the objective.
3. Answer the *Questions* at the end of the chapter in your text.
4. Complete the student study guide.

Key Terms/Notes

Define each of the following terms in the space provided, or make flash cards of the following terms.

Ecology

Consumers

Environment

Trophic level

Biotic factors

Herbivores

Abiotic factors

Primary consumers

Ecosystem

Carnivores

Producers

Secondary consumers

Primary carnivores

Biomass

Secondary carnivore

Community

Population

Food chain

Food web

Biomes

Omnivore

Decomposer

Climax community

Successional stage (community)

Sere

Primary succession

Pioneer community

Pioneer organisms

Secondary succession

Productivity

Biosphere

Questions with Short Answers

1. If the producer trophic level of an ecosystem contains 100 kg of biomass, the primary consumer level will have about _____ kg of biomass.

2. The rate at which an ecosystem accumulates new organic matter is known as its _____.

3. A biome characterized by the absence of trees and presence of permanently frozen soil is known as _____.

4. Temperate deciduous forest, desert, and savanna are examples of _____.

5. A _____ consists of all the interacting organisms in an area.

6. Herbivores occupy the _____ trophic level.

7. In a food chain consisting of trees, insects that eat the leaves of the trees, and birds that eat the insects, which group will have the largest biomass? _____.

8. _____ are animals that will eat a variety of food, some of which is plant material and some of which is animal material.

9. The branch of science which deals with the study of the way organisms interact with their environment is known as _____.

10. Nonliving organic matter is utilized by _____ as a source of energy.

11. Sunlight, length of day, air temperature, and kinds of seasons are all examples of _____ factors.

12. An _____ consists of a collection of organisms and the abiotic factors in an area.

13. A pyramid of energy will have _____ at its base.

14. A group of organisms of the same species is known as a _____.

15. The process of changing from one kind of community to another is called _____.

16. Succession that begins because the original community was destroyed is called _____ succession.

17. The first organisms to invade an area are known as _____ organisms.

18. A stable stage in succession that can maintain itself for a long period is known as the _____ community.

19. _____ that are too dry to allow for farming typically have been used as grazing land for cattle and sheep.

20. In most cases, _____ ecosystems contained fewer species and in some cases were entirely different kinds of communities from the originals.

Label/Diagram/Explain

Label the energy pyramid. Identify each trophic level and the general role-classification of the organisms in each of the 4 trophic levels. If there were 1000 kilocalories of energy at the first trophic level, how much would be present in the second, third, and fourth?

Energy Flow Through Ecosystems

Multiple Choice Questions

1. As energy is passed from one level to the next in a food chain about, _____% is lost at each transfer.
   A. 10
   B. 50
   C. 75
   D. 90

2. The primary factor that determines whether a geographic area will support temperate deciduous forest or prairie is
   A. the amount of rainfall.
   B. the severity of the winters.
   C. the depth of the soil.
   D. the kinds of animals present.

3. Which one of the following populations of organisms in an ecosystem would have the largest total biomass?
   A. herbivores
   B. producers
   C. carnivores
   D. decomposers

4. The concept of a community differs from that of an ecosystem in that
   A. a community includes plants and animals but not decomposers while an ecosystem includes all of these.
   B. a community does not include abiotic factors and an ecosystem does.
   C. a community is larger than an ecosystem.
   D. a community cannot be organized in a pyramid and an ecosystem can.

5. Which one of the following organisms is most likely to be at the second trophic level?
   A. a maple tree
   B. a snake
   C. a fungus
   D. an elephant

6. Which one of the following is typical for organisms in a food chain?
   A. most organisms will only use one species of organism as a source of food
   B. most organisms occupy 3–4 different trophic levels depending on what they happen to be eating at the time
   C. most organisms will be involved in several different food chains
   D. most organisms will always feed at the producer level

7. In secondary succession in forested areas annual weeds are replaced by grasses which are replaced by trees because
   A. larger plants shade smaller plants.
   B. more soil is produced.
   C. water becomes more abundant.
   D. animals eat the smaller plants.

8. Where would you find primary succession occurring?
   A. a clear-cut forest
   B. abandoned agricultural field
   C. prairie burned by fire
   D. rock exposed by glaciers

9. Tree ⟶ insect ⟶ spider ⟶ frog ⟶ fish ⟶ human
   In the food chain above, the spider is

   A. a tertiary consumer.
   B. a herbivore.
   C. at the third trophic level.
   D. a producer.

10. If the first trophic level contains 10,000 units of energy, the _____ trophic level will contain _____ units of energy.
    A. fifth; 10
    B. second; 9,000
    C. third; 100
    D. second; 5,000

11. Which of the following is biotic?
    A. water
    B. bacteria
    C. pH
    D. energy

12. The majority of the humans on Earth generally occupy which trophic level?
    A. first
    B. second
    C. third
    D. fourth

13. The biome you would typically find located between temperate deciduous forest and tundra is
    A. prairie.
    B. savanna.
    C. desert.
    D. boreal forest.

CHAPTER 15
COMMUNITY INTERACTIONS

Overview

Within ecosystems, organisms influence one another in many ways. Even organisms of the same species affect one another in the course of their normal daily activities. Many kinds of interactions occur within ecosystems and there are a variety of ways that organisms within communities affect each other in the cycling of matter.

Study Activities

1. Write a summary of each section of the *Chapter Outline* in your text.
2. For each of the *Learning Objectives* in your text, write a sentence or paragraph that demonstrates your mastery of the objective.
3. Answer the *Questions* at the end of the chapter in your text.
4. Complete the student study guide.

Key Terms/Notes

Define each of the following terms in the space provided, or make flash cards of the following terms.

Habitat   Host

Niche   External parasite

Predation   Internal parasite

Prey   Commensalism

Predator   Mutualism

Parasitism   Symbiosis

Parasite   Competition

Competitive exclusion principle   Denitrifying bacteria

Symbiotic and free-living nitrogen-fixing bacteria   Biological amplification

Nitrifying bacteria   Vector

Questions with Short Answers

1. The major cause of extinction or endangerment of species by humans is _____.

2. _____ bacteria live in the roots of plants and convert nitrogen gas into a form plants can use.

3. Plants release water to the atmosphere as a result of _____.

4. Animals obtain nitrogen in the _____ in the food they eat.

5. Two nitrogen-containing molecules that plants can use to obtain nitrogen are nitrates and _____.

6. In the carbon cycle, carbon atoms are released into the atmosphere as _____ gas.

7. The process by which carbon dioxide is removed from the atmosphere and the carbon atoms are incorporated into organic molecules is _____.

8. _____ is an interaction between organisms in which both are harmed.

9. A relationship between two organisms in which both benefit is _____.

10. The role of an organism is its _____.

11. The _____ is killed and eaten by a predator.

12. A _____ gets nourishment from a host.

13. Tapeworms are examples of _____ parasites.

14. Only _____ energy comes to the earth in a continuous stream.

15. It is important to recognize that although _____ results in harm to both organisms there can still be winners and losers.

Label/Diagram/Explain

Label the diagram below. Starting with carbon dioxide molecules in the atmosphere, discuss the processes that would allow the carbon atom to be passed from one organism to another until it is ultimately returned to the atmosphere in the form of a carbon dioxide molecule. Include in your discussion how each of the organisms in the diagram plays a different role in the community. Also discuss the two metabolic processes necessary for the carbon cycle.

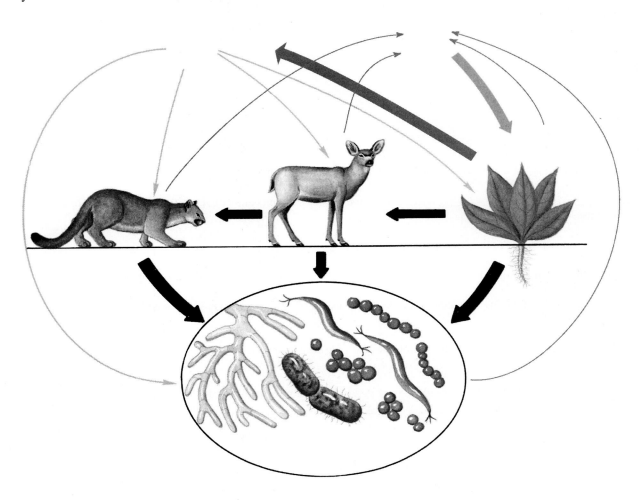

Multiple Choice Questions

1. Which one of the following describes, in part, the niche of a rabbit?
   A. the wind in the area it lives
   B. the golf course it lives on
   C. rabbits are eaten by coyotes
   D. sunlight

2. An epiphyte is in a _____ relationship.
   A. commensal
   B. parasitic
   C. competitive
   D. mutualistic

3. If two species of organisms occupy the same niche
   A. mutualism will result.
   B. competition will be very intense.
   C. both organisms will become extinct.
   D. both will need to enlarge their habitat.

4. Mutualism, parasitism, and commensalism are all examples of
   A. nitrogen-fixing bacteria.
   B. symbiosis.
   C. habitats.
   D. competition.

5. Water enters the atmosphere as a result of
   A. photosynthesis.
   B. respiration.
   C. evaporation.
   D. condensation.

6. If nitrogen-fixing bacteria were to become extinct
   A. life on earth would stop immediately since there would be no source of nitrogen.
   B. life on earth would continue indefinitely.
   C. life on earth would slowly dwindle as useful nitrogen became less available.
   D. life on earth would be unchanged except that proteins would be less common.

7. Nitrogen of the organic molecules in your body comes from
   A. the air you breath.
   B. the water you drink.
   C. carbohydrates in the food you eat.
   D. proteins in the food you eat.

8. Nitrogen is returned to the atmosphere as N_2 by
   A. plants.
   B. nitrogen-fixing bacteria.
   C. dentirifying bacteria.
   D. nitrifying bacteria.

9. If an ecosystem has been contaminated with DDT or PCBs, where would you find the highest concentrations of these chemicals?
   A. in the water
   B. in tissues of producers
   C. in fat tissues of primary consumers
   D. in fat tissues of secondary consumers

10. Nitrogen gas makes up approximately _____ of the Earth's atmosphere.
    A. 2%
    B. 7%
    C. 45%
    D. 80%

11. Which one of the following does NOT cycle through an ecosystem?
    A. energy
    B. water
    C. carbon
    D. nitrogen

12. The habitat of an earthworm is
    A. the planet Earth.
    B. the forest ecosystem.
    C. topsoil.
    D. eating dead organic matter.

13. Many plants (flowers) provide nectar for insects. The insects in turn pollinate the flower. This relationship between the insect and plant represents
    A. parasitism.
    B. commensalism.
    C. mutualism.
    D. predation.

14. Mosquitos do not cause malaria, but carry and transfer the organism that does cause malaria. Mosquitos in this instance are playing the role of a(n):
    A. vector.
    B. predator.
    C. epiphyte.
    D. competitor.

CHAPTER 16
POPULATION ECOLOGY

Overview

Populations of organisms show many kinds of characteristics. An introduction to some of these characteristics will help you see how populations grow and how this growth is controlled. The principles that apply to other populations of organisms can be related to the problems associated with the human population explosion.

Study Activities

1. Write a summary of each section of the *Chapter Outline* in your text.
2. For each of the *Learning Objectives* in your text, write a sentence or paragraph that demonstrates your mastery of the objective.
3. Answer the *Questions* at the end of the chapter in your text.
4. Complete the student study guide.

Key Terms/Notes

Define each of the following terms in the space provided, or make flash cards of the following terms.

Population

Gene flow

Gene frequency

Age distribution

Population density

Population pressure

Sex ratio

Reproductive capacity

Biotic potential

Population growth curve

Natality

Mortality

Lag phase

Exponential growth phase

Stationary growth phase

Extrinsic factors

Carrying capacity

Intrinsic factors

Limiting factors

Density-dependent factors

Environmental resistance

Density-independent factors

Death phase

Questions with Short Answers

1. A predator killing an organism would be an _____ limiting factor to a population.

2. If the intensity of a limiting factor increases as the size of the population increases, it is a _____ limiting factor.

3. The _____ is the optimum number of organisms of a species that can be supported over a long time.

4. The initial portion of a population growth curve in which the size of the population is growing slowly is the _____.

5. The period when population is growing most rapidly is known as the _____ phase.

6. The _____ is the theoretical maximum reproductive rate.

7. The _____ refers to the number of males in a population compared to the number of females.

8. As the number of individuals in a population increases the population _____ increases.

9. _____ is the number of individuals entering the population by birth.

10. As population density increases the rate of population growth will decrease because of _____ limiting factors.

11. The numbers of individuals in different age categories within a population is the _____ of the population.

12. All the limiting factors that act on a population to limit its size are collectively known as _____.

13. The current growth rate of the human population is _____.

14. The number of individuals leaving a population by death is called _____.

15. If a country has a total _____ rate of 2.1 children per woman the population of the country should stabilize.

Label/Diagram/Explain

Label the typical population growth curve below.

1. Place the LETTER A on the portion of the curve that indicates where natality is greater than mortality.
2. Place the LETTER B on the portion of the curve that indicates the population has reached its carrying capacity.
3. Place the LETTER C on the portion of the curve that indicates the current status of the human population growth rate.
4. Place the LETTER D on the portion of the curve that indicates that the population has mortality and natality equal.
5. Place the LETTER E on the portion of the curve that indicates the organism has a high biotic potential.
6. Place the LETTER F on the portion of the curve that indicates the population is just beginning to grow.

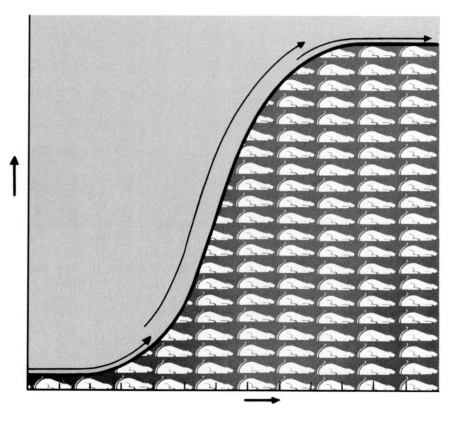

Multiple Choice Questions

1. As population density increases, which one of the following is likely to occur?
   A. natality will increase
   B. mortality will decrease
   C. the population will experience exponential growth
   D. individuals will migrate from the area of highest density

2. Density-independent limiting factors
   A. increase in intensity as the population increases.
   B. are unrelated to population size.
   C. usually influence the size of populations of large animals.
   D. never affect population size.

3. Which one of the following is NOT the direct result of increasing human population?
   A. extinction of some kinds of animals
   B. increased standard of living
   C. decreased availability of energy
   D. pollution

4. Populations of organisms that are in small, confined situations often have their population limited by
   A. the production of their own wastes.
   B. an inability to reproduce.
   C. reduced biotic potential.
   D. increased energy input.

5. A population made up primarily of prereproductive individuals will
   A. increase rapidly in the future.
   B. become extinct.
   C. rarely occur.
   D. remain stable for several generations.

6. Some organisms have a low reproductive capacity but are very successful because
   A. they have a short life span.
   B. they have high mortality.
   C. most of their offspring live.
   D. they do not compete with other organisms.

7. Which one of the following is an extrinsic limiting factor?
   A. fights between males over females
   B. competition over food
   C. death due to unusual weather
   D. increased sexual activity

8. Listed below are the sex ratios for four populations. All other things being equal (including current population size), which population should experience the greatest future growth?
   A. 1 male: 1 female
   B. 2 male: 1 female
   C. 1 male: 2 female
   D. 3 male: 2 female

9. Listed below are the natality and mortality numbers for four populations. All other things being equal, which population will experience the greatest growth?
   A. natality = 26/1000; mortality = 17/1000
   B. natality = 19/1000; mortality = 8/1000
   C. natality = 13/1000; mortality = 25/1000
   D. natality = 11/1000; mortality = 5/1000

10. Mortality exceeds natality during the _____ phase.
    A. death
    B. lag
    C. stable equilibrium
    D. exponential growth

11. An exponential increase:
    A. 100, 200, 300, 400, 500.
    B. 1, 2, 3, 4, 5.
    C. 2, 4, 8, 16, 32.
    D. 5, 10, 15, 20, 25.

12. The current human population is experiencing
    A. a population decline.
    B. slow steady growth.
    C. stable equilibrium.
    D. exponential growth.

13. Increasing a population's food supply may directly increase
    A. environmental resistance on the population.
    B. the carrying capacity for the population.
    C. the population's mortality.
    D. population pressure.

14. When yeasts ferment the sugar in grape juice, they produce ethyl alcohol. When alcohol concentration reaches a certain level, the yeast population declines and eventually dies. In this example, population growth of the yeast is stopped by
    A. limited space.
    B. limited food supply.
    C. accumulation of waste.
    D. disease.

Overview

Biologists have their own perspective on the various types of behaviors that enable organisms to survive, reproduce, and successfully compete in their environments. Some of these behaviors are simple while others are highly complex. As you read and study this material reflect on the previous information and attempt to integrate all that you have learned to better understand how the described behaviors enhance organisms' lives. Try your best not to "anthropomorphize," that is, attribute human feelings, meanings, and emotions to the behavior of the animals described. This will help you to NOT draw false conclusions.

Study Activities

1. Write a summary of each section of the *Chapter Outline* in your text.
2. For each of the *Learning Objectives* in your text, write a sentence or paragraph that demonstrates your mastery of the objective.
3. Answer the *Questions* at the end of the chapter in your text.
4. Complete the student study guide.

Key Terms/Notes

Define each of the following terms in the space provided, or make flash cards of the following terms.

Behavior

Operant Conditioning

Anthropomorphism

Classical conditioning

Ethology

Conditioned response

Instinctive behavior

Imprinting

Stimulus/Response

Insight learning

Learning

Pheromones

Territoriality

Dominance hierarchy

Territory

Photoperiod

Sign stimulus

Sociobiology

Intention movement

Society

Redirected aggression

Questions with Short Answers

1. The branch of science known as _____ deals with the study of the nature of behavior and its ecological and evolutionary significance.

2. The inherited behavioral "program" that gives an organism its best chance of survival is in the form of _____.

3. A special kind of behavior called _____ is based on past experiences that are reorganized to solve new problems.

4. Chemicals known as _____ are produced by some animals and released into the environment triggering behavioral or developmental changes in other animals of the same species.

5. Screaming at your spouse for some injustice your boss has imposed upon you is _____.

6. Many birds have their migration direction determined by the changing _____.

7. "Follow the leader" is a game that might be described as demonstrating _____.

8. The interactions among individual animals and the special roles each plays are characteristic of _____.

9. A _____ is a change that results in an altered behavior.

10. Getting upset when someone else sits in "your seat" is an example of _____.

11. At least a portion on some bird's songs is _____ behavior.

12. The drawback of _____ behavior is that it cannot be modified when a new situation presents itself.

Label/Diagram/Explain

Explain the following using information from the text:

1. Every time the zookeeper walks past the tropical fish exhibit, the fish come to the front of the glass and follow him as he passes.

2. "I refuse to eat strawberries because I know I'll break out in hives."

3. "Sit, Spot, sit!"

4. "Blow in his ear and he'll follow you anywhere!"

5. "This perfume is guaranteed to get you a date!"

6. ". . . like a bee flies to honey."

Multiple Choice Questions

1. Mate selection in animals often involves many behaviors including
   A. instinct.
   B. pheromones.
   C. learned behavior.
   D. all the above are possible.

2. Many animals learn to relate unpleasant experiences with color, taste, shape and other physical features. This is the result of
   A. associative learning.
   B. stimulus/response.
   C. photoperiodism.
   D. imprinting.

3. Behavior in which a young animal is genetically ready to learn a specific behavior is known as
   A. insight learning.
   B. imprinting.
   C. conditioned response.
   D. anthropomorphism.

4. There is evidence that some animals navigate by
   A. social behaviors.
   B. conditioned response.
   C. earth's magnetic poles.
   D. the aroma of food.

5. "That's a mouth watering aroma!"
   A. associative learning
   B. Pavlovian response
   C. classical conditioning
   D. all the above

6. Which of the following demonstrates the greatest amount of instinctive behavior?
   A. lady bugs
   B. flatworms
   C. birds
   D. mammals

7. The specific pattern of light flashes of fireflies is behavior that
   A. enables a population to respond to stimulation.
   B. allows a species to define it's territory.
   C. better insures reproduction of the species.
   D. is learned behavior.

8. Foxes have frequently been depicted in fables as sly and corrupt animals. This is an example of
   A. anthropomorphism.
   B. ethology.
   C. instinctive behavior.
   D. imprinting.

9. Two young boys have a disagreement on the playground. One of the boys stands up straight, sticks out his chest, and raises his fists as he stares sternly at the other boy. This posturing of the boy exemplifies
   A. sign stimuli.
   B. intention movements.
   C. redirected aggression.
   D. imprinting.

10. Your cat runs into the kitchen whenever it hears the sound of a can being opened. This happens regardless of whether the can is cat food, soup or coffee. This behavior of your cat is
    A. a conditioned response.
    B. imprinting.
    C. insight learning.
    D. instinctive behavior.

11. Your algebra teacher presents you with a problem you have never seen before. Based on other algebra problems you have solved in the past, you are able to work through the problem and come up with the correct solution. This is an example of
    A. associative learning.
    B. instinctive behavior.
    C. insight learning.
    D. conditioning.

12. Which of the following is an example of behavior?
    A. a duckling following its mother
    B. a plant bending towards light
    C. a bacterium dividing
    D. all of the above

13. Instinctive behaviors are the result of
    A. natural selection.
    B. conditioning.
    C. imprinting.
    D. insight learning.

CHAPTER 18
MATERIALS EXCHANGE IN THE BODY

Overview

The significance of cell surface area and cell volume must be appreciated in order to understand the movement of materials across cell surfaces. The transport of materials in and out of cells is an important concept to understand, whether the organism is single-celled or multicellular. In addition, multicellular animals must have special organs with large surface areas to accomplish materials exchange. Understanding the concept of surface-area-to-volume ratio is important to appreciate how the respiratory, circulatory, digestive, and excretory systems function.

Study Activities

1. Write a summary of each section of the *Chapter Outline* in your text.
2. For each of the *Learning Objectives* in your text, write a sentence or paragraph that demonstrates your mastery of the objective.
3. Answer the *Questions* at the end of the chapter in your text.
4. Complete the student study guide.

Key Terms/Notes

Define each of the following terms in the space provided, or make flash cards of the following terms.

Surface-area-to-volume ratio

Capillaries

Homeostasis

Lacteal

Heart

Pulmonary circulation

Aorta

Systemic circulation

Arteries

Lymphatic system

Veins

Immune system

Alveoli

Gall bladder

Diaphragm

Large intestine

Salivary gland

Villi

Pharynx

Kidneys

Duodenum

Nephrons

Hepatic portal vein

Bowman's capsule

Pancreas

Glomerulus

Bile

Coronary arteries

Questions with Short Answers

1. The main function of the large intestine is the _____ of water.

2. The _____ of the lungs are where gases are exchanged between air and the blood.

3. _____ are the blood vessels that exchange material between the blood and cells.

4. _____ are the functional units of the kidneys.

5. Blood vessels that carry blood from the heart are the _____.

6. Bile is produced in the _____.

7. The intercostals and the _____ are the muscles responsible for breathing.

8. The _____ is the section of the small intestines that receives secretions from the pancreas and liver.

9. Finger-like projections known as the _____ increase the surface area of the intestines.

10. A mass of capillaries in a kidney nephron is a _____.

11. An increase in the level of carbon dioxide in the blood will _____ the pH.

12. The blood that flows through the aorta to all parts of the body passes through the _____ circulation.

13. _____ is an iron-containing molecule that carries oxygen.

14. The _____ is the liquid portion of the blood.

15. The _____ is a large tube that carries air to the bronchi.

16. The _____ cells are capable of producing antibodies.

17. _____ can be found throughout the body and are the most active of the phagocytes.

18. Substances known as _____ are capable of stimulating an immune response.

Label/Diagram/Explain

1. Certain changes may occur in your body as a result of your environment or life-style. With this thought in mind answer the following.

    a. Explain why people who live at high elevations have a higher concentration of red blood cells and a higher hemoglobin level than people who live at lower elevations.

    b. Often heavy smokers, spray painters, and coal miners have difficulty in breathing. (In miners the condition is called black lung). Why do these people experience shortness of breath?

    c. If a person contracted a disease that interfered with their ability to produce carbonic anhydrase, explain the consequences that would result.

2. Label the following diagrams.

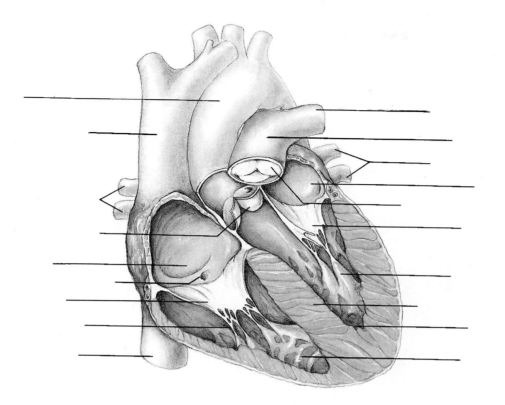

Fom John W. Hole, Jr., *Human Anatomy and Physiology,* 6th ed. Copyright © 1993 Times Mirror Higher Education Group, Inc., Dubuque, Iowa. All Rights Reserved. Reprinted by permission.

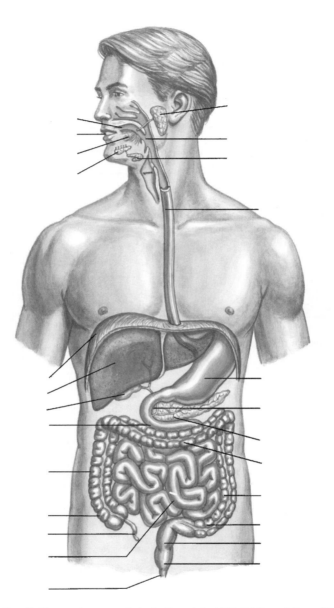

From Kent M. Van De Graaff, *Human Anatomy,* 3rd ed. Copyright 1992 © Times Mirror Higher Education Group, Inc., Dubuque, Iowa. All Rights Reserved. Reprinted by permission.

Multiple Choice Questions

1. Reabsorption of water occurs in association with
   A. lungs.
   B. duodenum.
   C. loop of Henle.
   D. liver.

2. The primary structures involved in pumping blood are the
   A. veins.
   B. atria.
   C. capillaries.
   D. ventricles.

3. Fats are broken down through the action of
   A. salivary amylase.
   B. bolus.
   C. pepsin.
   D. bile.

4. The fluid portion of the blood that leaves the capillaries and surrounds the cells is
   A. hemoglobin.
   B. edema.
   C. lymph.
   D. lacteal.

5. In the kidneys, glucose molecules are reabsorbed by the
   A. glomerulus.
   B. proximal convoluted tubule.
   C. distal convoluted tubule.
   D. microvilli.

6. Semilunar valves prevent the flow of blood into the
   A. aorta.
   B. atria.
   C. ventricles.
   D. arteries.

7. Which one of the following results in an increase in surface area?
   A. chewing
   B. reabsorption
   C. breathing
   D. salivation

8. Which cells are immune defenders and phagocytes?
   A. neutropohiles
   B. T-cells
   C. B-cells
   D. RBCs

9. Blood is carried through vessels to all parts of the body except the lungs by
   A. pulmonary circulation.
   B. the Pulmonary artery.
   C. systemic circulation.
   D. the lymphatic system.

10. Which of the following is FALSE concerning a cell that grows larger?
    A. its volume increases
    B. its metabolic needs increase
    C. it must exchange more materials with the environment
    D. its surface-area-to-volume ratio increases

11. A blood pressure reading (120/80 for example) is
    A. pressure while the heart is not contracting/pressure while heart is contracting.
    B. pressure while exercising/ pressure while at rest.
    C. systolic blood pressure/ diastolic blood pressure.
    D. diastolic blood pressure/ systolic blood pressure.

12. As air passes through the lungs it follows the path:
    A. trachea $\rightarrow$ bronchi $\rightarrow$ bronchioles $\rightarrow$ alveoli.
    B. trachea $\rightarrow$ bronchioles $\rightarrow$ bronchi $\rightarrow$ alveoli.
    C. bronchi $\rightarrow$ trachea $\rightarrow$ alveoli $\rightarrow$ bronchioles.
    D. bronchioles $\rightarrow$ alveoli $\rightarrow$ bronchi $\rightarrow$ trachea.

13. The structure of _____ provide an example the human body maximizing surface area for the exchange of materials.
    A. capillaries
    B. alveoli
    C. villi
    D. all of the above

14. Bile produced by the liver is stored in the
    A. duodenum.
    B. gallbladder.
    C. blood.
    D. pancreas.

15. The levels of water, hydrogen ions, salts and urea in the blood are regulated by the
    A. liver.
    B. kidneys.
    C. bladder.
    D. rectum.

CHAPTER 19
NUTRITION, FOOD, AND DIET

Overview

The biochemical pathways and physiological processes occur in humans as in all other living organisms. You are truly what you eat. The chemicals we call food undergo reactions that convert them into components of human cells, tissues, and organs. A better understanding of foods, dietary requirements, and how foods are processed can be helpful when selecting healthful combinations of foods.

Study Activities

1. Write a summary of each section of the *Chapter Outline* in your text.
2. For each of the *Learning Objectives* in your text, write a sentence or paragraph that demonstrates your mastery of the objective.
3. Answer the *Questions* at the end of the chapter in your text.
4. Complete the student study guide.

Key Terms/Notes

Define each of the following terms in the space provided, or make flash cards of the following terms.

Complete proteins

Incomplete proteins

Essential amino acids

Protein-sparing

Vitamins

Minerals

Osteoporosis

Electrolytes

Nutrients

Assimilation

Diet

Nutrition

Kilocalorie (kcalorie)

Calorie

Basal metabolic rate (BMR)   Kwashiorkor

Essential fatty acid   Specific dynamic action (SDA)

Recommended Dietary Allowances (RDAs)   Obese

Anorexia nervosa   Vitamin-deficiency disease

Four basic food groups   Carbohydrate loading

Questions with Short Answers

1. Water, vitamins, and minerals are three of the six classes of _____ required by the body.

2. _____ amino acids cannot be made in your body and must be taken in your diet.

3. Inorganic molecules needed in small amounts for proper nutrition are _____.

4. The basic food group that supplies most of the calcium is _____.

5. A person 30% above their ideal body weight is classified as _____.

6. During a person's life, their total energy requirements are the highest during _____.

7. Your _____ consists of your daily intake of nutrients.

8. Energy intake is measured in _____.

9. _____ is the most common nutrient in the body.

10. Your diet must furnish a supply of _____ in order for your digestive tract to function properly.

11. The _____ basic food group is a source of vitamin C.

12. Lactating women require an increase in their _____ intake.

13. People having a fear of being overweight may suffer from _____.

14. A lack of vitamins in the diet may lead to a vitamin _____ disease.

15. The last nutrient to be used as an energy source is _____.

16. _____ are dissolved inorganic ions.

Label/Diagram/Explain

1. It seems that many people are obsessed with losing weight. As a result there is an endless source of new diets, most with the promise of quick weight loss.

   You read about a carbohydrate diet that supplies only one thousand kcalories a day. The diet limits your nutrient intake to carbohydrates and water. The author of this diet suggests that you remain on it until you reach your desired weight goal. It is an inexpensive diet, carbohydrates are cheap; and with only one thousand kcalories a day, you should lose weight.

   Would this diet work? Why or why not? What are the advantages and disadvantages of such a diet?

2. Label the food guide pyramid and identify the number of servings of each food group that are recommended for a healthy adult.

Source: U.S. Department of Agriculture.

Multiple Choice Questions

1. Persons of normal weight who overeat and force the food out of their digestive tract
   A. are obese.
   B. are bulimic.
   C. have anorexia nervosa.
   D. have beriberi.

2. A complete protein contains
   A. all the Recommended Dietary Allowances.
   B. essential fatty acids.
   C. the suggested kcalorie amount.
   D. all the essential amino acids.

3. A person with kwashiorkor suffers from a lack of
   A. proteins.
   B. minerals.
   C. vitamins.
   D. kcalories.

4. Converting a nutrient into a molecule in your body's cells is
   A. digestion.
   B. assimilation.
   C. carbohydrate loading.
   D. protein-sparing.

5. Persons A and B have the same BMR. Person A weighs 30% more than Person B.
   A. person A eats more food
   B. person B eats more food
   C. both eat the same
   D. cannot tell from data presented

6. Which one of the following need to be most careful in developing a diet?
   A. pregnant woman
   B. athlete
   C. mature adult
   D. sickle cell anemia patient

7. RDAs are easily met by
   A. eating breakfast cereal.
   B. using vitamin supplements.
   C. eating a well balance, varied diet.
   D. consuming more fiber.

8. Carbohydrates that are not digested in the human digestive tract are
   A. fibers.
   B. minerals.
   C. vitamins.
   D. fatty acids.

9. A calorie is the energy required to raise one _____ of water one degree _____.
   A. gram; Fahrenheit
   B. cup; Fahrenheit
   C. gram; Celsius
   D. kilogram; Celsius

10. The measure of the amount of energy a person uses while at rest is
    A. BMR.
    B. SDA.
    C. diet.
    D. RDA.

11. The Calories or kilocalories listed on package labels are equal to _____ calories.
    A. 1
    B. 100
    C. 1 000
    D. 10 000

12. The amount of energy you burn each day is determined by your
    A. activity level.
    B. age and gender.
    C. body size.
    D. all of the above.

13. Osteoporosis results from a _____ deficiency.
    A. vitamin
    B. calcium
    C. protein
    D. energy

14. Most Americans should have more _____ in their diet.
    A. protein
    B. fat
    C. sugar
    D. complex carbohydrates

15. Without dietary planning, _____ is (are) most likely to be missing from a vegetarian diet.
    A. complex carbohydrates
    B. essential fatty acids
    C. fiber
    D. essential amino acids

CHAPTER 20
THE BODY'S CONTROL MECHANISMS

Overview
The control of various body functions is important to the survival of an organism. Because many different systems each perform a specific set of tasks in the human body, it is necessary to coordinate these various activities with one another. This coordination is done by the nervous system and the endocrine system. These systems are also responsible for various types of sensory input and the response mechanisms that are involved in the control of the body.

Study Activities
1. Write a summary of each section of the *Chapter Outline* in your text.
2. For each of the *Learning Objectives* in your text, write a sentence or paragraph that demonstrates your mastery of the objective.
3. Answer the *Questions* at the end of the chapter in your text.
4. Complete the student study guide.

Key Terms/Notes
Define each of the following terms in the space provided, or make flash cards of the following terms.

Stimulus

Soma

Response

Axon

Nervous system

Dendrite

Endocrine system

Central nervous system

Glands

Peripheral nervous system

Neuron

Nerve impulse

Synapse

Tympanum

Acetylcholine

Semicircular canals

Hormones

Actin

Negative-feedback control

Myosin

Perception

Motor unit

Retina

Endocrine glands

Rods

Exocrine glands

Cones

Fovea centralis

Questions with Short Answers

1. A _____ is a group of muscle cells stimulated by a neuron.

2. In muscles thin filaments of actin alternate with thick filaments of _____.

3. _____ are chemical messengers secreted by the endocrine glands.

4. A change in the environment that an organism can detect is a _____.

5. In a nerve cell, the _____ carries impulses away from the cell body.

6. In the retina, the _____ detect colored light.

7. The nervous system acts on two types of organs, _____ and muscles.

8. The brain and the spinal cord are parts of the _____ nervous system.

9. Secretions produced by the _____ glands are released directly into the circulatory system.

10. The tympanum is stimulated by _____ changes in the environment.

11. A series of wave depolarizations in a neuron results in a _____.

12. A _____ is the recognition by the brain that a stimulus has been received.

13. The space between the axon of one neuron and the dendrite of another neuron is a _____.

14. In the eye, the _____ is that portion that is sensitive to light.

15. The neurotransmitter that functions at a synapse is _____.

16. The _____ system operates more slowly than the nervous system.

17. _____ ions are pumped out from a cell by active transport.

Label/Diagram/Explain

Use the illustration below to explain the polarization of cell membranes. Next, draw and label a nerve cell. Explain the transmission of an impulse in a neuron and explain how this impulse is transmitted from one neuron to another neuron.

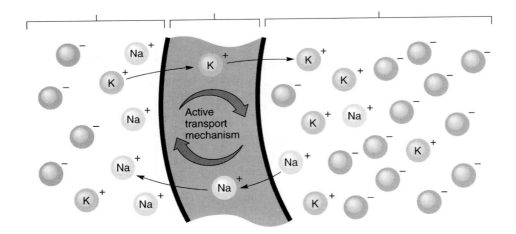

Multiple Choice Questions

1. An organism's reaction to a change in the environment is a(n)
   A. stimulus.
   B. impulse.
   C. response.
   D. perception.

2. At night which one of the following is used?
   A. rods
   B. green cones
   C. red cones
   D. cochlea

3. When muscles contract
   A. glycogen is synthesized.
   B. calcium is lost.
   C. ATP is produced.
   D. actin slides by myosin.

4. Acetylcholine is destroyed by
   A. cholinesterase.
   B. endocrine glands.
   C. exocrine glands.
   D. axons.

5. The source of growth stimulating hormone responsible for the disease Giantism is
   A. adrenal gland.
   B. pituitary gland.
   C. testes.
   D. thyroid.

6. A light stimulus is received by the nervous system which results in growth. This is the result of
   A. release of hormones from the thyroid.
   B. activating muscles.
   C. stimulating the endocrine system.
   D. increasing nervous activity.

7. When a nerve cell is stimulated
   A. acetylcholine is destroyed.
   B. potassium ions enter the neuron.
   C. sodium ions enter the neuron.
   D. calcium attaches to the dendrites.

8. Which one of the following is necessary for muscle contraction?
   A. calcium ions
   B. glucose
   C. testosterone
   D. fat

9. When the temperature of a home falls below a set-point, the furnace produces heat. Once the home has reached the desired temperature, the furnace shuts off. As the home cools and again falls below the set-point the furnace will again produce heat. This cycle is similar to _____ that occurs in the human body.
   A. depolarization
   B. negative feedback control
   C. muscle contractions
   D. a synapse

10. Which one of the following is NOT a type of muscle?
    A. skeletal muscle
    B. cardiac muscle
    C. neuron muscle
    D. smooth muscle

11. The central nervous system consists of the
    A. brain only.
    B. brain and spinal cord.
    C. brain, spinal cord, and nerves.
    D. motor neurons and sensory neurons.

12. Olfactory senses detect
    A. light.
    B. sound.
    C. chemicals.
    D. pain.

13. The ear bones are the
    A. malleus, incus, and stapes.
    B. tympanum and cochlea.
    C. rods and cones.
    D. fovea centralis and olfactory epithelium.

14. The chemical messengers secreted by endocrine glands are called
    A. enzymes.
    B. hormones.
    C. neurotransmitters.
    D. impulses.

15. Which of the following is NOT part of a nerve cell?
    A. dendrites
    B. soma
    C. axon
    D. myosin

CHAPTER 21
HUMAN REPRODUCTION, SEX, AND SEXUALITY

Overview
Sex and our sexuality influences us in many different ways throughout life. Before birth, sex-determining chromosomes direct the formation of hormones that control the development of sex organs, after which their effects diminish. The power of our sexuality and the reproductive urge demand that the production of offspring be considered an important aspect of our nature. At different stages in life, sex and sexuality play different roles.

Study Activities
1. Write a summary of each section of the *Chapter Outline* in your text.
2. For each of the *Learning Objectives* in your text, write a sentence or paragraph that demonstrates your mastery of the objective.
3. Answer the *Questions* at the end of the chapter in your text.
4. Complete the student study guide.

Key Terms/Notes
Define each of the following terms in the space provided, or make flash cards of the following terms.

Sexuality

Hormones

Autosomes

Puberty

Sex-determining chromosome

Hypothalamus

X chromosome

Pituitary gland

Y chromosome

Follicle-stimulating hormone

Differentiation

Estrogen

Conception

Androgens

Secondary sexcharacteristics

Corpus luteum

Menstrual cycle

Sperm

Ovulation

Vagina

Interstitial-cell-stimulating-hormone

Semen

Testosterone

Oogenesis

Gametogenesis

Polar body

Spermatogenesis

Follicle

Testes

Zygote

Cryptorchidism

Placenta

Orgasm

Menopause

Barr body   Gonads

Ovaries   Uterus

Masturbation   Penis

Ejaculation   Inguinal hernia

Questions with Short Answers

1.   1.   The two kinds of sex-determining chromosomes are _____ and _____.

2.   Chromosomes that do not determine the sex of and individual are called _____.

3.   A female has two _____ chromosomes.

4.   The process of cells becoming specialized for certain functions is called _____.

5.   _____ are molecules that control the differentiation of cells and organisms.

6.   The period of time when an individual completes their sexual development to become sexually mature is called _____.

7.   Characteristics that differentiate males from females but are not directly related to reproduction are called _____ sexual characteristics.

8.   In females the pituitary gland secretes _____ hormone which stimulates the ovaries to produce estrogen.

9.   The periodic growth and shedding of the lining of the uterus is known as the _____ cycle.

10. _____ is the regular release of a sex cell from the ovary.

11. The primary male sex hormone is _____.

12. The series of cell divisions and maturation processes that results in the production of sperm is called _____.

13. During oogenesis the cellular division is unequal resulting in the formation of _____ bodies.

14. The release of sperm and seminal fluid is known as _____.

15. The single cell that results from the union of an egg and sperm is a _____.

16. The _____ allows for the exchange of molecules between mother and embryo.

94

Label/Diagram/Explain

Label the following diagram of gametogenesis. Do **NOT** look at the diagram in your text while doing this.

Gametogenesis

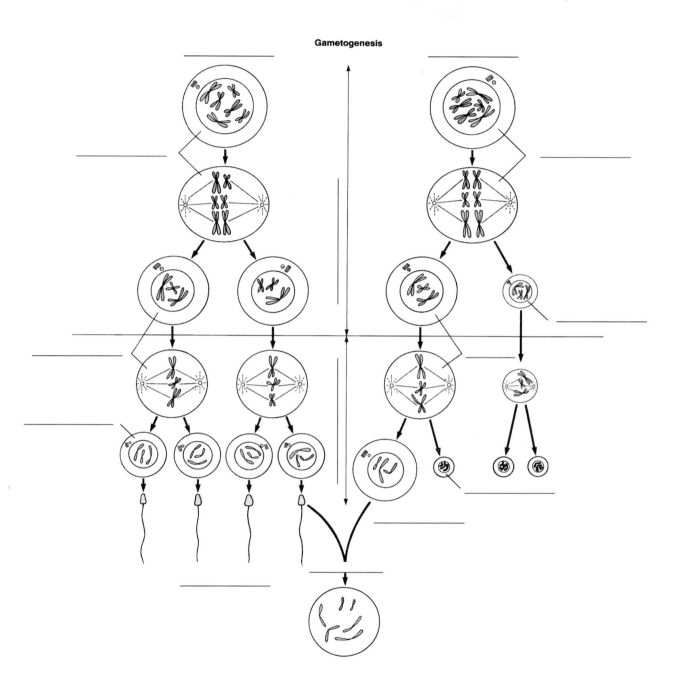

Multiple Choice Questions

1. You could identify human cells as being either male or female by
   A. looking at the size of a cell.
   B. examining the chromosomes.
   C. looking for differentiation.
   D. weighing them.

2. Ovaries and testes differ from other kinds of organs because
   A. their cells divide.
   B. their cells only live for a short time.
   C. meiosis occurs in them.
   D. they grow continuously.

3. The sex of an individual is set
   A. at conception.
   B. at birth.
   C. at puberty.
   D. during pregnancy.

4. Which one of the following would be a secondary sex characteristic?
   A. ovaries in a female
   B. the presence of a penis in a male
   C. a menstrual cycle in a female
   D. pubic hair in males and females

5. Ovulation is most likely to occur
   A. during the middle of the menstrual cycle.
   B. after the menstrual cycle.
   C. any time during the menstrual cycle.
   D. before the first meiotic division.

6. Which one of the following hormones is responsible for stimulating the testes to produce testosterone?
   A. interstitial-cell-stimulating-hormone
   B. estrogen
   C. androgens
   D. oxytocin

7. Which one of the following is NOT a difference between oogenesis and spermatogenesis?
   A. spermatogenesis doesn't involve meiosis
   B. spermatogenesis produces 4 sperm and oogenesis produces one viable sex cell
   C. sperm are smaller than eggs
   D. many more sperm are produced than eggs

8. Which of the following contraceptive methods is most effective?
   A. condom
   B. diaphragm
   C. pill
   D. spermicides

9. In a healthy adult male, each ejaculation will release approximately _____ sperm.
   A. 300
   B. 3000
   C. 3 million
   D. 300 million

10. In humans, oogenesis begins
    A. before birth.
    B. at puberty.
    C. each month.
    D. at menopause.

11. Fraternal twins
    A. result from the fertilization of two different oocytes by two different sperm.
    B. result when a zygote splits into two separate cells which develop independently.
    C. are always the same sex.
    D. are genetically identical.

12. A woman with Turner's Syndrome would have the sex chromosome combination:
    A. XX.
    B. XY.
    C. XO.
    D. XXY.

13. The condition in which a portion of the intestines push through the inguinal canal into the scrotum is called
    A. an inguinal hernia.
    B. cryptorchidism.
    C. a vasectomy.
    D. a cesarean.

CHAPTER 22
THE ORIGIN OF LIFE AND EVOLUTION OF CELLS

Overview

Consideration of the origin of life gives you the opportunity to draw together knowledge gained previously to explore one of the most fascinating hypothetical topics. While many have proposed interesting ideas over the past centuries, it has only been within the last century that biochemical, cellular anatomy and physiology, and astrophysics have contributed to the body of knowledge surrounding this work.

Study Activities

1. Write a summary of each section of the *Chapter Outline* in your text.
2. For each of the *Learning Objectives* in your text, write a sentence or paragraph that demonstrates your mastery of the objective.
3. Answer the *Questions* at the end of the chapter in your text.
4. Complete the student study guide.

Key Terms/Notes

Define each of the following terms in the space provided, or make flash cards of the following terms.

Spontaneous generation

Heterotrophs

Biogenesis

Autotrophs

Reducing atmosphere

Oxidizing atmosphere

Prebionts

Prokaryote

Coacervate

Eukaryote

Microsphere

Endosymbiotic theory

Proteinoids

Archaea

Questions with Short Answers

1. The concept that all forms of life only spring from already existing life is referred to as _____.

2. The French scientist, _____, convinced many others of his time that spontaneous generation did not happen.

3. According to Oparin and Haldane, the first organic molecules were formed in an early _____ atmosphere.

4. The _____ concept received a great deal of support in 1996 when a meteorite from Antarctica was analyzed.

5. Some believe the first cells functioned as _____, i.e., cells that require complex organic molecules from their environment.

6. _____ are cells that are able to photosynthesize.

7. Types of cells that possess nuclear membranes are called _____.

8. An atmosphere that is predominantly H_2, CH_4, NH_3, and H_2O, with no O_2 is _____.

9. The ultimate source of energy for all living things on Earth today is the _____.

10. According to the endosymbiotic theory, _____ were in early times free living nonphotosynthetic bacteria.

11. Scientists have also shown that _____ molecules are able to make copies of themselves without the need for enzymes.

12. The molecule _____ found in the upper atmosphere helps to screen out UV light.

13. The Age of the _____ lasted longer than the Age of the Dinosaurs.

14. O_2 is characteristic of an _____ atmosphere.

15. Many kinds of very primitive prokaryotic organisms known as the _____ are autotrophic and live in extremely hostile environments.

16. It is thought that Earth is at least _____ years old.

17. _____ published the idea that the "bacteria" are a group of similar organisms and are really made up of two very different kinds of organisms; the Bacteria and Archaea.

Label/Diagram/Explain

Label the diagram below and explain the Endosymbiotic Theory.

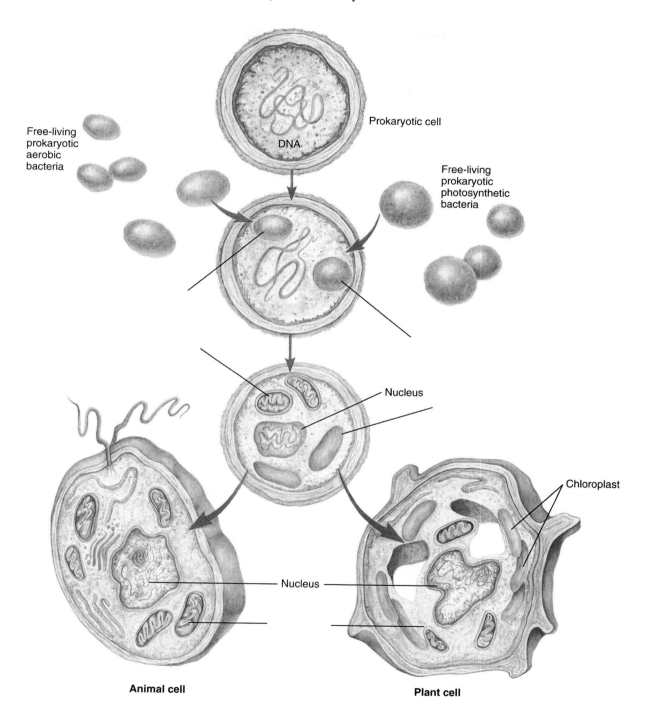

Prokaryotic cell

Free-living prokaryotic aerobic bacteria

DNA

Free-living prokaryotic photosynthetic bacteria

Nucleus

Chloroplast

Nucleus

Animal cell

Plant cell

Multiple Choice Questions

1. Which was NOT a major component of the early Earth's reducing atmosphere?
   A. H_2
   B. CO_2
   C. NH_3
   D. ozone

2. Hypothetically, which came first in the evolution of cells?
   A. autotrophs
   B. heterotrophs
   C. aerobic cells
   D. eukaryotes

3. Prokaryotic cells hypothetically came into existence approximately _____ years ago?
   A. 20 billion
   B. 4-5 billion
   C. 3.5 billion
   D. 1.5 billion

4. According to the Endosymbiotic Theory
   A. chloroplasts were cyanobacteria.
   B. mitochondria were cyanobacteria.
   C. the original cell type was eukaryotic.
   D. microspheres were nuclei.

5. Which was an early source of energy for chemical reactions in the early Earth's atmosphere?
   A. volcanos
   B. ATP
   C. X-rays
   D. all of these

6. This theory proposes that the solar system was formed from a large cloud of gases.
   A. Endosymbiotic
   B. Solar Nebula
   C. Little Bang
   D. Prebiont

7. Which of the following is a macromolecule?
   A. H_2
   B. prokaryote
   C. ATP
   D. Ozone

8. Which one of the following researchers supported spontaneous generation?
   A. Redi
   B. Needham
   C. Spallanzani
   D. Pasteur

9. The three main kinds of living things Bacteria, Archaea, and Eucarya have been labeled as
   A. Species.
   B. Kingdoms.
   C. Domains.
   D. Families.

10. Miller's apparatus
    A. simulated conditions that may have existed on early Earth.
    B. demonstrated spontaneous generation of cells in a reducing atmosphere.
    C. supported the theory of biogenesis.
    D. contradicted the earlier work of Pasteur.

11. Which event or step was NOT necessary to produce life from inorganic materials?
    A. Organic molecules must be formed from inorganic molecules.
    B. The inorganic molecules must be collected together and segregated from other molecules by a cell wall.
    C. Collections of organic molecules must become self-sustaining by making new molecules as older ones are randomly destroyed.
    D. Ultimately this first cellular unit must be able to reproduce more of itself.

12. The first cells probably had higher mutation rates than cells today because
    A. RNA served as the genetic material.
    B. there was no protective ozone in the atmosphere.
    C. the first life forms were unstable.
    D. nucleotides had not yet evolved.

13. The theory that proposes how the first eukaryotic cells evolved is
    A. biogenesis.
    B. spontaneous generation.
    C. the endosymbiotic theory.
    D. the solar nebula theory.

CHAPTER 23
THE CLASSIFICATION AND EVOLUTION OF ORGANISMS

Overview

The great diversity and number of organisms found on planet Earth are mind boggling. In order to relieve some of the confusion and help focus one's attention when studying any particular type of organism, scientists have developed a classification system. This system is based on hypothetical evolutionary relationships among organisms but sometimes organisms just don't fit our system. As we learn more about organisms, their position in the classification system is changed to reflect our more detailed level of understanding. This can cause some confusion in itself, but the intent is to make things easier to understand and more accurate.

Study Activities

1. Write a summary of each section of the *Chapter Outline* in your text.
2. For each of the *Learning Objectives* in your text, write a sentence or paragraph that demonstrates your mastery of the objective.
3. Answer the *Questions* at the end of the chapter in your text.
4. Complete the student study guide.

Key Terms/Notes

Define each of the following terms in the space provided, or make flash cards of the following terms.

Taxonomy

Prion

Binomial system of nomenclature

Alternation of generations

Phylogeny

Virus

Saprophyte

Obligate intracellular parasite

Fungus

Host

Gametophyte

Pandemic

Questions with Short Answers

1. In the scientific name, *Streptococcus pyogenes,* the name _____ is the species name.

2. The common edible mushroom is a member of the kingdom _____.

3. The Swedish doctor and botanist _____ originated the modern system of taxonomy.

4. Some organisms grow into free-living developmental stages called _____ that do not resemble the adults of their species.

5. A common method of cellular reproduction among the bacteria _____, consists of DNA replication and cytoplasmic division.

6. Members of the Kingdom _____ lack a true nucleus.

7. The protozoans and algae are placed in the Kingdom _____.

8. Organisms that cause the diseases valley fever, "ringworm", and athlete's foot are commonly called _____.

9. The _____ is the haploid stage in a plant's life cycle that produces sex cells.

10. It is thought that the _____ evolved from the flagellated protista.

11. A _____ is a protein coated nucleic acid particle that functions as an obligate intracellular parasite.

12. The _____ is a specific kind of cell the provides what the virus needs to replicate.

13. The _____ is the taxonomic subdivision immediately under the kingdom level of classification.

14. _____ are organisms that feed on dead organisms.

15. The study of an organism's evolutionary history is the science of _____.

16. A _____ is a division of a phylum.

Label/Diagram/Explain

Each of us is familiar with many kinds of organisms from our local environment. Make a list of the five kingdoms and give at least three examples of organisms found in each. Using these organisms, arrange them in an evolutionary "bush".

Multiple Choice Questions

1. Which of the following scientific names is written correctly?
   A. *Micropterus* Salmoides
   B. *Treponema pallidum*
   C. Nymphaea odorata
   D. salmo *Trutta*

2. Which is the proper sequence of taxa?
   A. kingdom, phylum, class, order, family
   B. phylum, kingdom, family, order, class
   C. phylum, class, order, family, species
   D. kingdom, phylum, family, class, order

3. Which is commonly used as sources of information when developing a phylogeny?
   A. fossil evidence
   B. biochemical information
   C. DNA analysis
   D. all the above are used

4. Which type of organism is NOT included in the five kingdom system of classification?
   A. fungi
   B. marine red algae
   C. viruses
   D. larva

5. The single-cell Mycetae are commonly called
   A. algae.
   B. molds.
   C. protozoa.
   D. yeast.

6. Which of the following Kingdoms contain members that are autotrophs?
   A. Mycetae
   B. Animalia
   C. viruses
   D. Protista

7. Infectious diseases that occur throughout the world at unacceptably high rates are referred to as
   A. endemic.
   B. epidemic.
   C. pandemic.
   D. international.

8. The cause of AIDS is a
   A. bacterium.
   B. fungus.
   C. protozoan.
   D. virus.

9. Of the following, two organisms which belong to the same _____ are most closely related.
   A. family
   B. order
   C. class
   D. kingdom

10. Alternation of generations is a characteristic of
    A. Prokaryotae.
    B. Protista.
    C. Mycetae.
    D. Plantae.

11. Viruses
    A. can be free-living or parasitic.
    B. belong to the kingdom Prokaryotae.
    C. can only function and reproduce inside a host.
    D. all of the above.

12. An organism that is multicellular, eukaryotic and heterotrophic is a(n)
    A. plant.
    B. animal.
    C. protist.
    D. animal or fungus.

13. Organisms that obtain energy from the decomposition of other organisms are
    A. saprophytes.
    B. gametophytes.
    C. sporophytes.
    D. obligate intracellular parasites.

14. Most Plantae are
    A. cone bearing plants.
    B. ferns.
    C. flowering plants.
    D. mosses.

15. Algae, amoeba, and paramecia belong to thekingdom
    A. Mycetae.
    B. Prokaryotae.
    C. Protista.
    D. virus.

16. Viruses are composed of
    A. prokaryotic cells.
    B. eukaryotic cells.
    C. protein and nucleic acid.
    D. membranous organelles.

CHAPTER 24
PROKARYOTAE, PROTISTA, AND MYCETAE

Overview

Many people are unfamiliar with the members of these three kingdoms unless they have some special reason to become aware of these microorganisms. That might mean they develop a disease caused by one of these microbes, have to buy one for use in their home or business, or read about an incident that relates to them. While most members of these kingdoms are microscopic, some are enormous. The reason for referring to them as microbes is based on their cellular anatomy and physiology, evolutionary relationships, and intercellular relationships. It has been said that there are more microbes in a teaspoon of soil than humans on the face of the Earth. That should be justification alone for learning more about this group.

Study Activities

1. Write a summary of each section of the *Chapter Outline* in your text.
2. For each of the *Learning Objectives* in your text, write a sentence or paragraph that demonstrates your mastery of the objective.
3. Answer the *Questions* at the end of the chapter in your text.
4. Complete the student study guide.

Key Terms/Notes

Define each of the following terms in the space provided, or make flash cards of the following terms.

Microorganisms

Protozoa

Microbes

Algae

Colonial

Plankton

Bacteria

Benthic

Pathogens

Phytoplankton

Endospores

Bloom

Mycorrhiza   Gram stain

Mycotoxins   Protista

Mycetae   Archaeobacteria

Questions with Short Answers

1. A _____ is a collection of microbes of the same species that live together in a typical arrangement and show a small degree of cooperation.

2. The _____ bacteria are responsible for supplying plants and other organisms with usable forms of nitrogen.

3. A microbe which is capable of causing disease is referred to as a _____.

4. The bacterium _____ is capable of forming highly resistant endospores.

5. The _____ are autotrophic, unicellular organisms.

6. _____ organisms live on the bottom of marine and freshwater habitats.

7. It is estimated that _____% of atmospheric O_2 is produced by phytoplankton.

8. The diatoms are algae that have _____ in their cell walls.

9. A sudden, explosive increase in the number of algae is called a _____.

10. Photosynthesis is the process whereby _____ make O_2 and carbohydrates using sunlight as the source of energy while growing in the ocean.

11. The protozoans are classified according their method of _____.

12. A _____ is a symbiotic relationship between some fungi and the roots of some plants.

13. A _____ is a symbiotic relationship between a fungus and an alga.

14. Toadstools produce _____.

15. The _____ have cell walls, ribosomes, cell membranes unlike those of eubacteria and are typically found in extreme environments.

Label/Diagram/Explain

While many people associate the term MICROBES with disease and environmental problems, microorganisms play many beneficial roles in the environment and our lives. Review the chapter and make a list of the helpful prokaryotes, protistans, and mycetes; and note the nature of their benefits.

Example: Mycetae: Fungus — produces the antibiotic penicillin.

Multiple Choice Questions

1. Fire blight and citrus canker are plant diseases caused by
   A. bacteria.
   B. fungi.
   C. protozoa.
   D. viruses.

2. Which of the following is required to destroy bacterial endospores?
   A. boiling water
   B. sun light
   C. 121°C, for 15-20 minutes under pressure
   D. 100° C for 10 minutes

3. Which of the following is a fungus-like protistan?
   A. slime mold
   B. diatom
   C. red algae
   D. althele's foot

4. Photosynthesis can occur in which zone?
   A. upper portion of a lake
   B. in the neritic zone
   C. photic zone
   D. all these are correct

5. Which is NOT a product of algae?
   A. alginate
   B. carrageenin
   C. agar
   D. mycotoxin

6. The vector of malaria is
   A. a ciliated protozoan.
   B. the mosquito, *Anopheles*.
   C. a Sporozoan.
   D. the common housefly.

7. The first chemical used against plant diseases was a copper-based fungicide called
   A. DDT.
   B. penicillin.
   C. Bordeaux mixture.
   D. silver phosphate.

8. Corn smut, fairy rings, and toadstools are all
   A. fungi.
   B. algae.
   C. bacteria.
   D. protozoa.

9. Protista differ from Prokaryotae in that Protista
   A. are unicellular.
   B. can be autotrophs or heterotrophs.
   C. can have cell walls.
   D. are composed of eukaryotic cells.

10. Pathogenic bacteria cause
    A. malaria and amoebic dysentery.
    B. ringworm and athlete's foot.
    C. strep throat, syphilis, and pneumonia.
    D. all of the above.

11. Which of the following involves a mutualistic relationship?
    A. lichen
    B. mycorrhizae
    C. methanogens
    D. all of the above

12. Which one of the following is a protistan?
    A. slime mold
    B. yeast
    C. moss
    D. cyanobacteria

13. Small organisms which float and move with water currents are
    A. endospores.
    B. benthic.
    C. plankton.
    D. colonial.

14. Heterotrophic organisms can be found in the kingdom
    A. Prokaryotae.
    B. Protista.
    C. Mycatae.
    D. all of the above.

15. Mushrooms can produce _____ of spores.
    A. hundreds
    B. thousands
    C. millions
    D. billions

16. _____ produced by mushrooms are poisonous and/or hallucinogenic to humans.
    A. Antibiotics
    B. Alcohols
    C. Mycorrhizae
    D. Mycotoxins

CHAPTER 25
PLANTAE

Overview

Members of this kingdom are among the most familiar. They are all rather large and green. They are also among the most important life forms on Earth since, like the algae, they photosynthesize releasing O_2 into the atmosphere as they produce carbohydrates which heterotrophs may use as nutrients. Botanists continue to discover many new species of plants and learn more about their niche.

Study Activities

1. Write a summary of each section of the *Chapter Outline* in your text.
2. For each of the *Learning Objectives* in your text, write a sentence or paragraph that demonstrates your mastery of the objective.
3. Answer the *Questions* at the end of the chapter in your text.
4. Complete the student study guide.

Key Terms/Notes

Define each of the following terms in the space provided, or make flash cards of the following terms.

Flower

Germinate

Mosses

Alternation of generations

Gametophyte generation

Tissues

Antheridium/archegonium

Vascular tissue

Sporophyte/gametophyte

Xylem/phloem

Life cycle

Roots/root hairs

Capsule

Stems/leaves/roots

Seed

Cones

Cambium/wood

Pollen

Gymnosperms/angiosperms

Nondeciduous

Perennials/annuals

Monocots/dicots

Pistil/stamen

Imperfect/perfect flower

Accessory structures

Petals/sepals

Questions with Short Answers

1. There are three types of bryophytes: _____ mosses, liverworts, and hornworts.

2. The _____ is the stage in the life cycle of a plant which is responsible for the production of either sperm or egg.

3. In order to survive on land, plants needed special _____ tissues to conduct water from the soil to the aerial portions of the organism.

4. _____ is a series of hollow cells arranged end to end and conduct water to the upper portion of the plant.

5. Structures known as _____ increase the efficiency of absorbing water from the soil.

6. The portion of a fern that one sees when they buy a fern is really the

_____ generation of the organism.

7. In gymnosperms, two types of _____ are formed: male and female.

8. Male gymnosperms are responsible for the production of miniaturized male gametophytes called _____.

9. The first classification of plants was done in the fourth century by one of Aristotle's students, _____.

10. The xylem of a woody stem make up most of what we commonly refer to as

_____.

11. The ovary and other tissues of an angiosperm mature into a protective structure known as a

_____.

12. The seed leaves of an angiosperm as called _____.

110

13. Flowers with both male and female reproductive
    structures are

    _____.

Label/Diagram/Explain

Label the following diagrams:

Figure 1: Flower Parts

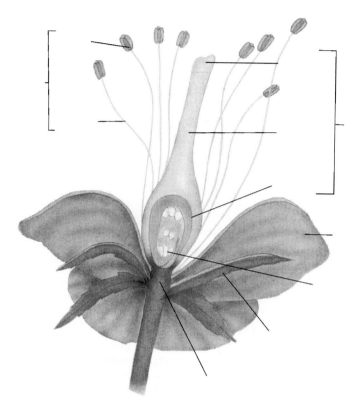

Figure 2: Life Cycle of a Fern

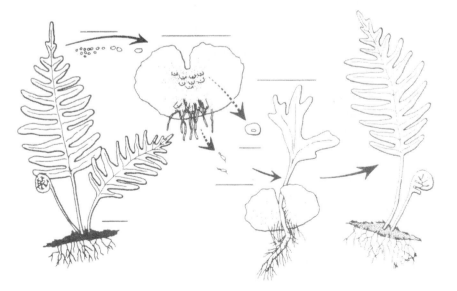

Multiple Choice Questions

1. The filament of a moss is called a
   A. bud.
   B. protonema.
   C. xylem.
   D. sporophyte.

2. Which of the following does NOT have vascular tissue?
   A. angiosperms
   B. gymnosperms
   C. ferns
   D. liverworts

3. The vascular tissue that conducts organic molecules from the upper portion of the plant to the roots.
   A. protonema
   B. xylem
   C. phloem
   D. cambium

4. These structures of angiosperms and gymnosperms contain an embryonic plant.
   A. spores
   B. seeds
   C. pollen
   D. all of these may be in a seed

5. These kinds of plants loose their leaves each year.
   A. deciduous
   B. pollination
   C. germination
   D. imperfect

6. The outer, dead, rough covering of some vascular plants is known as
   A. cambium
   B. wood
   C. bark
   D. xylem

7. Peaches, apples, acorns, and tomatoes are all examples of
   A. fruits.
   B. flowers.
   C. sepals.
   D. dicots.

8. Grass is a
   A. monocot.
   B. dicot.
   C. gymnosperm.
   D. fern.

9. A plant with many primitive characteristics is a
   A. fern.
   B. gymnosperm.
   C. moss.
   D. angiosperm.

10. A pine tree is a(n)
    A. monocot.
    B. dicot.
    C. gymnosperm.
    D. angiosperm.

11. The dominant structures of which of these plants is haploid?
    A. fern
    B. gymnosperm
    C. moss
    D. angiosperm

12. Which of the following is NOT part of the pistil?
    A. style
    B. anther
    C. ovary
    D. stigma

13. A seed contains a(n)
    A. ovary.
    B. gamete.
    C. embryo.
    D. gametophyte.

14. A plant's first leaf is called a
    A. sepal.
    B. cotyledon.
    C. pistil.
    D. xylem.

15. Pollination
    A. produces a male gamete.
    B. is the transfer of pollen from the male reproductive structure to the female reproductive structure.
    C. is the union of egg and sperm.
    D. is the emergence of a seedling.

CHAPTER 26
ANIMALIA

Overview

This is it! The end! The top of the heap! . . . Or that's what most humans believe! Don't forget that without plants, members of the Kingdom Animalia would not exist. As with plants, animals can be found in both aquatic and terrestrial environments. They have evolved into a great variety of species that occupy a variety of ecological niches. While humans usually refer to "the animals" as if they we were not a member of this large diverse group, remember that all human beings are classified in the Kingdom Animalia along with animals that vary from starfish to elephants.

Study Activities

1. Write a summary of each section of the *Chapter Outline* in your text.
2. For each of the *Learning Objectives* in your text, write a sentence or paragraph that demonstrates your mastery of the objective.
3. Answer the *Questions* at the end of the chapter in your text.
4. Complete the student study guide.

Key Terms/Notes

Define each of the following terms in the space provided, or make flash cards of the following terms.

Poikilotherm   Mesenteries

Homeotherm   Sessile

Asymmetry   Filter feeders

Radial symmetry   Vertebrates/invertebrates

Bilateral symmetry   Budding

Coelom/acoelomates   Medusa/polyp

Benthic   Hibernation/Torpor

Pelagic   Geological Time Chart

Questions with Short Answers

1. In most animals, the majority of cells are located _____ the organism.

2. Animals whose body temperature varies with their metabolism and the environment are known as _____.

3. The _____ is the cavity inside many animals in which internal organs are suspended.

4. Animals that spend much of their lives attached to a portion of their environment are said to be _____.

5. Many sponges demonstrate a _____ body plan that is constructed around a central axis.

6. The _____ are dorsal bony supports of animals such as birds and mammals.

7. The majority (99.9%) of animals on the Earth today lack a dorsal supporting structure and are known as _____.

8. The free-swimming adult stage, _____, of many marine animals is the location of sexual reproductive organs.

9. *Schistosoma mansoni,* a member of the group of animals know as the _____, is responsible for infections of the human circulatory system.

10. The _____ are characterized by a soft body enclosed by a hard shell.

11. The water vascular system is a unique feature of animals commonly called _____ even though they are not true fish.

12. Because most fish swim freely as adults they are called _____.

13. Sharks and rays differ from perch and red snapper in that the sharks and rays have a skeleton composed of _____.

14. The jointed legged _____ are among the most populace animal groups on Earth.

15. _____ occurs when an animal enters a *long term, seasonal* condition of hypothermia (lowered body temperature), ex. squirrels, bats, and badgers.

Label/Diagram/Explain

List the primary features of the following animal groups. Check the features of each that enable them to survive on land.

Arthropods:

Amphibians:

Reptiles:

Birds:

Mammals:

Multiple Choice Questions

1. Which of the following mammals is an egg layer?
   A. whale
   B. platypus
   C. armadillo
   D. kangaroo

2. Which of the following is an amniotic egg producer?
   A. amphibian
   B. grasshopper
   C. snake
   D. fish

3. A tube system found in insects that allows for the exchange of CO_2 and O_2 between the animal's body and the environment.
   A. trachea
   B. lungs
   C. Malphigian tubules
   D. allantois

4. Which is NOT true of the arthropods?
   A. they have separate sexes
   B. they have chitinous exoskeleton
   C. they have compound eyes
   D. they are homeothermic

5. Which does not belong in this list?
   A. tadpole
   B. salamander
   C. toad
   D. turtle

6. Which change contributed to the evolution of animals onto the land?
   A. internal fertilization
   B. the shelled egg
   C. water tight skin
   D. all the above

7. Which are the two major taxonomic categories of fish?
   A. bony and cartilaginous
   B. star and bony
   C. marine and freshwater
   D. sharks and rays

8. Which of the following animals has radial symmetry?
   A. earthworm
   B. turtle
   C. snail
   D. jellyfish

9. Most animals are
   A. vertebrates.
   B. homeotherms.
   C. asymmetrical.
   D. poikilotherms.

10. Sessile organisms
    A. are attached to substrate — stay in one place.
    B. are generally filter feeders.
    C. usually have free-swimming larval stages.
    D. all the above.

11. Which does not belong in this list?
    A. lobster
    B. beetle
    C. snail
    D. millipede

12. The first animals to live on land found the environment favorable because _____ on land than in water.
    A. it was easier to move
    B. temperatures varied less
    C. gas exchange (O_2 and CO_2) was easier
    D. there were many more unfilled niches

13. Tapeworms are
    A. flatworms.
    B. roundworms.
    C. segmented worms.
    D. earthworms.

14. A body cavity which separates the gut from the outer body is called a
    A. coelom.
    B. mesentery.
    C. medusa.
    D. polyp.

15. The geological period when fishes dominate the seas, the first insects and the first amphibians move onto land.
    A. Devonian
    B. Carboniferous
    C. Silurian
    D. Ordovician

Concepts in Biology 9th ed.
Student Study Guide Answers

CHAPTER 1 - WHAT IS BIOLOGY?

Questions with Short Answers

1. Biology
2. Nonscience
3. Responsive
4. Generative process
5. Cells
6. Uptake
7. Tissue
8. Theoretical/basic
9. Hypothesis
10. Empirical

Label/Diagram/Explain

Observation: The researchers have noticed that their plants have become tall and spindly.

Question formulation: The people who work in the laboratory wonder why the plants grow this way. The formal question is: "What factors determine that plants will grow rapidly enough to become tall and spindly?"

Explore alternative resources: The research team read information that had been reported in scientific journals indicating that one possible reason for the rapid growth was low light intensities. Chances are the rest of the team spent some time discussing the problem among themselves and with others and read a variety of reports. They had probably checked a bibliography or a bibliographic service, but had come up with only the report about low light as a causative agent.

Hypothesis formation: After discussing a wide variety of possibilities, they think that the reason the plants grow so tall and spindly is that lower light intensities stimulate plants to produce more stem tissue and, therefore, more leaf tissue in order to expose itself to as much light as possible. They think that a reasonable guess is that lower light intensities cause more rapid (spindly growth) whereas higher light intensities cause shorted plants.

Experimentation: They need to do a controlled experiment to determine if in fact light is the causative agent. Therefore, they establish two groups of plants: one set of 50 cuttings from one plant in the greenhouse, another set of 50 cuttings from the same plant in a low light greenhouse. Results showed that low light plants (experimental group) grew taller than high light (control group) plants. The two groups of plants were treated exactly the same. They were from the same plant so they had the same genetic environment, they were watered exactly the same, they were fertilized exactly the same, and the atmospheres were the same. Any differences in growth pattern had to be due to the light intensity differences.

Publication and peer review: They published their data and the laboratories repeated the experiments. Although others may not have gotten exactly the same results, the two sets of data were not significantly different. Therefore, the first laboratory is even more sure that its hypothesis is correct.

Multiple Choice Questions

1. A
2. A
3. D
4. A
5. D
6. B
7. B
8 D
9. C
10. C
11. A
12. D
13. C
14. D
15. D

CHAPTER 2 - SIMPLE THINGS OF LIFE

Questions with Short Answers

1. Atom
2. Energy
3. Protons
4. Nucleus
5. Electrons
6. Neutral
7. Neutrons
8. Covalent
9. Ions
10. Hydrogen
11. Solid
12. Gas
13. Suspension
14. Reaction
15. Radioactive
16. Kinetic energy
17. Electrons
18. pH
19. Molecule
20. Polar molecules

Label/Diagram/Explain

1. This isotope has 19 protons, 19 electrons, and 21 neutrons. The mass of the isotope is 40 AMU (19 protons + 21 neutrons). Electrons have almost no mass and are not considered in this calculation.
2. The neutrons and the protons in this and all atoms are located in the nucleus of the atom. There are 19 protons and 21 neutrons in the nucleus.
3. Since this is an atom of potassium it has an equal number of positive and negative charges, 19 positively charged protons and, therefore, 19 negatively charged electrons.
4. Potassium is in the firs column of the periodic table of the elements. This means that it forms a positive ion by releasing one electron. When the one electron is lost, the remaining unit is

composed of 19 positively charged protons, but only 18 negatively charged electrons. The extra positive charge makes it a positive ion, a cation.
5. The resultant charge of the ion is positive or +1.
6. 7 electrons.

Multiple Choice Questions

1.	B	6.	B	11.	A
2.	D	7.	B	12.	B
3.	B	8.	D	13.	C
4.	A	9.	C	14.	B
5.	C	10.	B	15.	D

CHAPTER 3 - ORGANIC CHEMISTRY: THE CHEMISTRY OF LIFE

Questions with Short Answers

1.	Carbon	11.	Saturated
2.	Four	12.	Amino acids
3.	Covalent	13.	Peptide
4.	Structure	14.	Double
5.	Carbon skeleton	15.	Three
6.	Functional	16.	Alcohol
7.	Carbohydrates	17.	Polymer
8.	Oxygen	18.	Monosaccharide
9.	Carbohydrates	19.	Amno acids
10.	Fatty acids		

Label/Diagram/Explain

1. See text page 0000
2. See text figure 3.8.
3. Dipeptide; see text page 0000; dehydration synthesis
4. Triglyceride; one fatty acid would be replaced by a phosphate group on a phospholipid molecule; see text figure 3.12.
5. See text pages 0000 and 0000; the reactants are pentose sugars; the products are a disaccharide ($C_{10} H_{18} O_9$) and water ($H_2 O$); dehydration synthesis.

Multiple Choice Questions

1.	D	6.	A	11.	B
2.	A	7.	B	12.	D
3.	D	8.	D	13.	C
4.	A	9.	C	14.	D
5.	C	10.	D	15.	B

CHAPTER 4 - CELL HISTORY, STRUCTURE, AND FUNCTION

Key Terms/Notes

See text figure 4.11.

Questions with Short Answers

1.	Microtublues	9.	Chloroplasts
2.	Phospholipid	10.	Carbohydrate (food)
3.	Nucleoplasm	11.	Ribosomes
4.	Diffusion	12.	Spindle
5.	Active transport	13.	Cilia
6.	Organelles	14.	Eukaryotic
7.	Golgi	15.	Vacuole
8.	Respiration		

Label/Diagram/Explain

1. See text figure 4.5.
   Hypertonic solution-cell will take in water and swell.

   Hypertonic solution-cell will lose water and shrink.

   Isotonic solution-water will diffuse in and out of the cell equally (no change)

2. a. See text figure 4.7.
   b. Hydrophilic ends outside, hydrophobic ends inside
   c. Facilitated diffusion moves molecules with the concentration gradient (left to right)
   d. Active transport moves molecule against the concentration gradient (right to left)
   e. Active transport, ATP

Multiple Choice Questions

1.	C	6.	C	11.	D
2.	B	7.	D	12.	B
3.	D	8.	C	13.	A
4.	D	9.	D	14.	B
5.	B	10.	B		

CHAPTER 5 - ENZYMES

Questions with Short Answers

1. Attachment site
2. Activation
3. Catalyst
4. Enzyme
5. Coenzymes
6. Enzyme-substrate complex
7. Shape
8. Denature
9. Shape
10. Turnover number
11. Increase
12. Inhibition
13. Gene regulator
14. Enzymatic competition
15. Lower
16. Substrat
17. -ase

Label/Diagram/Explain

1. See text figure 5.2. During an enzyme-controlled reaction, the enzyme and substrate come together to form a new molecule—the enzyme—substrate complex molecule. This molecule exists for only a very short time. During that time, activation energy is lowered and bonds are changed. The result is the formation of a new molecule or molecules called the end products. The enzyme comes out of the reaction intact and ready to be used again.

2. See text figure 5.1. The top line on this graph represents the energy needed for the reaction to occur without the assistance of an enzyme. The lower line represents the energy needed for the reaction to occur with an enzyme. Enzymes operate by lowering the amount of energy needed to get a reaction going (activation energy).

3. See text figure 5.5. Small temperature changes would cause minor changes in the rate of movement of molecules. Increasing the temperature would cause the enzymes and substrate molecules to move more rapidly, resulting in more collisions. This increase in collisions would result in a greater number of effective collisions, so more of the substrate molecules would become product molecules. Small decreases in the temperature would result in a slowing of the motion and, therefore, a decrease in the number of collisions. A large temperature increase might very well cause the enzyme molecule (a protein) to become denatured. The enzyme would be changed so drastically that it could not attach to the substrate molecule, therefore, it could no longer function as an enzyme. This is an irreversible change, so even when the temperature reverts to normal, the protein is unable to serve as an enzyme. A large temperature decrease does not result in denatured enzymes, only more drastic slowing in the reaction rate.

4. See text figure 5.6. Changes in pH may cause changes in the shape of the enzyme. This might happen because some of the side chains of the individual amino acids are positively charged and some are negatively charged. (Some are also neutral in charge.) In high acid environments, the H^+ ions would crowd around the negative side chains and could cause distortions in the overall shape of the protein. In highly basic environments, the OH^- ions would crowd around the positive side chains and could cause distortions in the overall shape of the protein. These distortions could be great enough to cause the protein to be unable to fit the substrate, or if drastic enough, the changes in the pH could denature the protein.

5. As the concentration of the enzyme increases the time it takes for all of the substrate molecules to be changed would decrease, since there are more enzymes working. The reverse is true if the concentration of the enzyme is deceased. As the substrate is increased, more product would be produced since the enzymes would continue to change substrate into product. If substrate concentrations are decreased, the amount of product formed will decrease.

6. Inhibitor molecules interfere with the ability of the enzyme to form a complex with the substrate. As long as the inhibitor molecules occupy the active site, the enzyme is unable to attach to the substrate molecule and convert it to product. When the inhibitor is removed, the enzyme is again able to junction. It has not been denatured.

Multiple Choice Questions

1.	C	6.	D	11.	B
2.	B	7.	A	12.	A
3.	D	8.	C	13.	D
4.	D	9.	B	14.	C
5.	C	10.	A		

CHAPTER 6 -BIOCHEMICAL PATHWAYS

Questions with Short Answers

1. (Sun)light
2. Oxygen
3. Carbon dioxide
4. Chloroplast
5. Aerobic
6. ATP
7. Glycolysis
8. Mitochondria
9. Hydrogen/ Carbondioxide
10. Three
11. Enzymes
12. Eating
13. Electron transfer system
14. Fats
15. Pyruvic acid
16. Alcohol
17. Bacterial or muscle

Label/Diagram/Explain

1. L
2. K
3. C, G
4. L, C
5. G
6. K
7. C
8. C
9. G
10. L
11. L
12. E
13. G
14. G, K, E
15. G
16. L
17. C
18. K, E
19. L
20. G, K

21. Carbon dioxide is in low concentration in the atmosphere. It would enter the leaf through openings called stomates. It moves around between the cells of the leaf, and eventually diffuses into a cell, which contains chromoplasts and would enter the chloroplast. Carbon dioxide is a raw material for the carbon dioxide conversion stage of photosynthesis. The carbon dioxide ultimately becomes a part of an organic molecule such as PGAL or glucose. Either of these molecules can be sent to a neighboring mitochondrion, where they would be processed through aerobic cellular respiration to release their chemical bond energy to form ATP molecules. This process involves glycolysis, the Krebs cycle, and the ETS. During the Krebs cycle, carbon dioxide molecules and pairs of hydrogen atoms are removed. The hydrogen atoms are transferred to the ETS by hydrogen carriers (NAD^+ and FAD). Energy is provided in the form of ATP via the ETS, the carbon dioxide molecules are released into the spaces between the cells, and are then released into the atmosphere. Of course some of the carbon dioxide molecules might be used over again by a neighboring chloroplast; others simply diffuse into the environment.

Multiple Choice Questions

1.	C	6.	B	11.	B
2.	B	7.	C	12.	C
3.	A	8.	B	13.	A
4.	D	9.	D	14.	B
5.	A	10.	C		

CHAPTER 7 - DNA AND RNA: THE MOLECULAR BASIS OF HEREDITY

Questions with Short Answers

1. DNA
2. Cytosine
3. Sugar
4. Hydrogen
5. Guanine
6. Coding
7. RNA
8. DNA
9. Protein
10. Amino acid
11. Anticodon
12. Nucleus
13. Mutations
14. Genetic engineering
15. PCR
16. Introns
17. Jumping genes or transposons

Label/Diagram/Explain

1. See text figure 7.13.
2. See text figure 7.2.

a.	DNA	l.	Cystine
b.	RNA	m.	Guanine
c.	Nucleotides	n.	Adenine
d.	Phosphate	o.	Uracil
e.	Deoxyribose	p.	Phosphate
f.	Cytosine	q.	Sugar phosphate
g.	Guanine	r.	Hydrogen
h.	Adenine	s.	Adenine
i.	Thymine	t.	Thymine
j.	Phosphate	u.	Cytosine
k.	Ribose	v.	Guanine

3. a. Ribosome h. Codons
    b. Translation i. Peptide
    c. mRNA j. Dehydration synthesis
    d. tRNA k. Amino acid
    e. Amino acids l. mRNA
    f. Anticodons m. tRNA
    g. Hydrogen n. Water
4. See text figure 7.12.

Multiple Choice Questions

1.	B	6.	D	11.	C
2.	D	7.	C	12.	C
3.	B	8.	D	13.	A
4.	B	9.	B		
5.	A	10.	D		

CHAPTER 8 - MITOSIS: THE CELL-COPYING PROCESS

Questions with Short Answers

1.	Metaphase	8.	Centromere
2.	Chromatids	9.	Nuclear
3.	Spindle	10.	Cytokinesis
4.	Cleavage	11.	Metastasis
5.	Differentiation	12.	Interphase
6.	Cancer	13.	Apoptosis
7.	Daughter		

Label/Diagram/Explain

See text figure 8.10.

Multiple Choice Questions

1.	A	6.	B	11.	B
2.	D	7.	C	12.	D
3.	B	8.	C	13.	C
4.	D	9.	B	14.	A
5.	A	10.	D		

CHAPTER 9 - MEIOSIS: SEX-CELL FORMATION

Questions with Short Answers

1.	Crossing-over	9.	Telophase
2.	Independent assortment	10.	Prophase I
3.	Telophase I	11.	Prophase I
4.	Mitosis	12.	Variation
5.	Male	13.	Anaphase II
6.	Diploid	14.	Gonads
7.	Testes	15.	Trisomy
8.	First	16.	Nondisjunction

Label/Diagram/Explain

1. Prophase of mitosis, prophase I and prophase II of meiosis:
    Individual chromosomes become visible
    In some cells the centrioles duplicate and the spindle begins to form
    Nuclear membrane disintegrates
    Nucleoli disappear.

    Metaphase of mitosis, metaphase I, and metaphase II of meiosis:
    Chromosomes align at the equatorial plane
    Nuclear membrane and nucleoli not present
    Spindle formation completed and chromosomes attached

    Anaphase of mitosis, anaphase I, and anaphase II of meiosis:
    Nuclear membrane and nucleoli not present
    Spindle fibers exted from pole to pole
    Chromosomes move to the poles.

    Telophase of mitosis, telophase I, and telophase II of meiosis:
    Nuclear membranes reform
    Nucleoli reappear
    Cytokinesis splits the cytoplasm
    Spindle disappears.

2. Mitosis involves one division cycle; meiosis involves two; synapsis of homologous chromosomes occurs during meiosis I.

    Crossing-over occurs during meiosis I:

    Chromosome numbers are reduced by having Homologous chromosomes separate during anaphase I.

    Meiosis results in diploid cells being reduced to haploid cells, while mitosis can occur to either haploid or diploid cells and the chromosome numbers remain the same.

3. See text figures 9.9 and 9.14.

Multiple Choice Questions

1.	A	6.	D	11.	D
2.	B	7.	C	12.	A
3.	D	8.	A	13.	B
4.	C	9.	B	14.	C
5.	C	10.	A		

CHAPTER 10 - MENDELIAN GENETICS

Questions with Short Answers

1.	Phenotype	9.	Locus
2.	Homozygous	10.	Environmental factors
3.	Codominance	11.	Heterozygous
4.	1/1296	12.	Double-factor
5.	Allele	13.	SRY
6.	Heterozygous	14.	X-linked
7.	Polygenic inheritance		
8.	Sex-linked		

Label/Diagram/Explain

Explain to grandmother that individuals have two alleles for each trait; they receive one allele from each parent. (Don't worry about X and Y chromosomes).

Tell her that there are different traits involved: hair shape, hair color, and eye color, and that each is inherited independently. For example, hair color and hair shape are not inherited as a unit. Point out to her that some of Fred's children have curly, red hair an some have straight, red hair.

Tell her that some alleles are dominant and can hide the presence of recessive alleles. In your family most individuals have at least one allele for each of the following: curly hair, black hair, and brown eyes.

Explain that straight hair and red hair are recessive traits. Also explain that there are multiple alleles for hair color, and that eye color is determined by the combined influence of at least two genes.

Since Fred has children with straight hair, red hair, and green eyes, his wife must have the alleles for straight hair, red hair, and green eyes.

Since he has children with curly hair, black hair, and brown eyes, Sarah must also have the alleles for curly hair, black hair, and brown eyes, and would have the

same appearance (phenotype) as the majority of the members of the family.

Multiple Choice Questions

1.	D	6.	B	11.	D
2.	A	7.	C	12.	A
3.	D	8.	C	13.	B
4.	B	9.	C		
5.	B	10.	C		

CHAPTER 11 - DIVERSITY WITHIN SPECIES

Questions with Short Answers

1.	Species	9.	Population
2.	Gene pool	10.	Mutations or migration
3.	Clones	11.	Sexual
4.	Eugenic laws	12.	Species
5.	Hybrids	13.	Genotype
6.	Deme	14.	Deme
7.	Monoculture	15.	Genetic diversity
8.	Genetic counselor	16.	Species

Label/Diagram/Explain

This color is probably due to a mutation. It could be a recessive allele that has been in the breed for generations but has not appeared until two dogs heterozygous for the trait were bred. It could also be a mutation resulting in a dominant allele.

Cross your unique colored male with several unrelated females. It is highly probable these females lack the unusual coat color allele. If none of the offspring from these liters shows the unique color, the allele is almost certainly recessive. If half the pups have unique color, it is dominant. (Your male would be heterozygous for the color in that case.)

If the unique color is dominant, breed your male with females (probably his daughters) who have the desired color. Since both parents are probably heterozygous, about 1/4 of their offspring should be homozygous for the unique dominant coat-color trait. Select male offspring with the desired coat color and breed with many females. If one of the males only has offspring with the unique coat color, he is homozygous. You can then breed him to any female and all the offspring have the desired coat color, although most of hem will be heterozygous. You car in business.

If the unique coat color is recessive, breed the male with his mother, who must have the recessive allele, or his sisters, who may be carriers of the desired coat-color allele. Save all offspring that have the desired characteristic and use them as breeding stock.

If you neuter all the dogs you sell, you are the only breeder who can have unique colored dogs.

Multiple Choice Questions

1.	B	6.	D	11.	D
2.	C	7.	D	12.	B
3.	A	8.	D	13.	B
4.	C	9.	C		
5.	D	10.	A		

CHAPTER 12 -NATURAL SELECTION AND EVOLUTION

Questions with Short Answers

1.	Mutations	8.	Higher (greater)
2.	Genome	9.	More
3.	Acquired	10.	Mutation, migration
4.	Recombination	11.	Evolution
5.	Selecting agent	12.	Small, large
6.	Hardy-Weinberg Law	13.	Genetics
7.	Differential reproduction	14.	Genetic information

Label/Diagram/Explain

1. The insecticide is acting as a selecting agent. Those individual insects that are susceptible to the insecticide die and those that are resistant live to reproduce and pass their genes for resistance on to the next generation. The ladybird beetles are also acting as selecting agents. Some of the aphids may have characteristics that make them less susceptible to predation while others may be easy prey. Those not selected as prey will have more offspring and pass their genes on to the next generation of aphids.

2. The aphids will have the least genetic variety became they reproduce asexually. All individuals descended from one female will have exactly the same genes.

3. After using the same insecticide for several generations, there would have been continual elimination of those individuals who had genes that made them susceptible to the insecticide and selection for those genes that gave the aphids resistance to the insecticide. Therefore, fewer aphids would be killed by the insecticide.

4. The broccoli plants are being subjected to aphids as a selecting agent. Some plants may have genes that allow them to resist attack from aphids. (Most plants produce toxic materials that

discourage insects from eating them.) At the same time, insects have ways of counteracting the effects of the toxins. This is a never-ending process.

Multiple Choice Questions

1.	D	6.	B	11.	D
2.	C	7.	D	12.	D
3.	A	8.	C		
4.	B	9.	B		
5.	C	10.	C		

CHAPTER 13 - SPECIATION AND EVOLUTIONARY CHANGE

Questions with Short Answers

1.	Punctuated equilibrium	9.	Gene flow
2.	Convergent evolution	10.	Polyploidy
3.	Adaptive radiation	11.	Gradualism
4.	*Australophithicus*	12.	Neandertals
5.	Geographic barriers or geographic isolation	13.	*Homo sapiens*
6.	Subspecies		
7.	Species		
8.	Genetic (reproductive) isolating		

Label/Diagram/Explain

See text figure 13.3. In most cases the first step in the process of speciation would be *the establishment of geographic barriers* that would prevent the movement of individuals from one part of the range to other parts of the range. Once a local population has been isolated from other populations of the species it cannot exchange genes with other populations through sexual reproduction. Since the local environment from the isolated population is likely to be somewhat different from that of other local populations, *genes that produce characteristics that are adaptive to the local environment will be favored* and the local isolated population may begin to diverge from its parent population. *You would know that they are separate species when they will not naturally interbreed with other local populations even if the physical barriers are removed.* Geographic variants may look or behave differently, but when given the opportunity, will still be able to interbreed and produce fertile offspring.

Multiple Choice Questions

1.	A	5.	C	9.	B
2.	B	6.	B	10.	D
3.	B	7.	C	11.	A
4.	D	8.	C	12.	A

CHAPTER 14 - ECOSYSTEM ORGANIZATION AND ENERGY FLOW

Questions with Short Answers

1.	Ten	11.	Abiotic
2.	Productivity	12.	Ecosystem
3.	Tundra	13.	Producers (plants)
4.	Biomes	14.	Population
5.	Community	15.	Succession
6.	Second	16.	Secondary
7.	Trees	17.	Pioneer
8.	Omnivores	18.	Climax
9.	Ecology	19.	Grasslands
10.	Decomposers	20.	Successional

Label/Diagram/Explain

See text figure 14.5. About 90% of the energy is lost at each transfer from one trophic level to the next. Second trophic level - 100; third -10; fourth -1.

Multiple Choice Questions

1.	D	6.	C	11.	B
2.	A	7.	A	12.	B
3.	B	8.	D	13.	D
4.	B	9.	C		
5.	D	10.	C		

CHAPTER 15 - COMMUNITY INTERACTIONS

Questions with Short Answers

1.	Habitat destruction	8.	Competition
2.	Symbiotic nitrogen-fixing	9.	Mutualism
		10.	Niche
3.	Transpiration	11.	Prey
4.	Protein	12.	Parasite
5.	Ammonia	13.	Internal
6.	Carbon dioxide	14.	Sunlight
7.	Photosynthesis	15.	Competition

Label/Diagram/Explain

See text figure 15.9. Carbon dioxide is taken from the air by *producers* (plants and algae) and the carbon is incorporated into organic molecules by the *process of photosynthesis*. Once the carbon is present in the organic molecules of plants or algae it can be passed to herbivores when the *herbivores* eat plants. The organic molecules are modified and incorporated into the bodies of herbivores, which can be eaten by *carnivores*. When organisms die or produce waste products, these carbon-containing organic molecules are decomposed by various *decomposers*. All organisms, producers, herbivores, carnivores and decomposers carry on the *process of respiration* that breaks down organic material to water and carbon dioxide. The carbon dioxide is released into the atmosphere and the cycle is complete.

Multiple Choice Questions

1.	C	6.	C	11.	A
2.	A	7.	D	12.	C
3.	B	8.	C	13.	C
4.	B	9.	D	14.	A
5.	C	10.	D		

CHAPTER 16 - POPULATION ECOLOGY

Questions with Short Answers

1.	Extrinsic	9.	Natality
2.	Density-dependent	10.	Intrinsic
3.	Carrying capacity	11.	Age distribution
4.	Lag phase	12.	Environmental resistance
5.	Exponential growth	13.	Large (high, rapid)
6.	Reproductive capacity (Biotic potential)	14.	Mortality
7.	Sex ratio	15.	Fertility
8.	Density		

Label/Diagram/Explain

See text figure 16.5.

Multiple Choice Questions

1.	D	6.	C	11.	C
2.	B	7.	C	12.	D
3.	B	8.	B	13.	B
4.	A	9.	B	14.	C
5.	A	10.	A		

CHAPTER 17 - BEHAVIORAL ECOLOGY

Questions with Short Answers

1.	Ethology	7.	Dominance hierarchy
2.	DNA	8.	Societies
3.	Insight learning	9.	Stimulus
4.	Pheromones	10.	Territoriality
5.	Redirected aggression	11.	Learned
6.	Photoperiod	12.	Instinctive

Label/Diagram/Explain

1. This is in all likelihood the result of a conditioned response related to the fact that the zookeeper's repeated presence may be associated with feeding of the fish.
2. This behavior is insight learning. The person may have previously had a rash following the eating of strawberries, and relate the result (rash) to the eating of the strawberries.
3. If the dog was rewarded each time it "sat" when given the command. Insight learning might also be involved in this situation.
4. Depending on the reward, this, too, could be classical conditioning. Insight learning might also be involved in this situation.
5. Again, this could be classical conditioning, but it may also be insight learning. Many events in our lives are commonly associated with aromas. If you smell a steak on the grill, your mouth might water. Similarly, we might associate sex appeal with specific odors.
6. Instinctive behavior. Even though the behavior of bees can be very complex, nearly all aspects of their behavior are instinctive, encoded in their genes.

Multiple Choice Questions

1.	D	6.	B	11.	C
2.	A	7.	C	12.	D
3.	B	8.	A	13.	A
4.	C	9.	B		
5.	D	10.	A		

CHAPTER 18 - MATERIALS EXCHANGE IN THE BODY

Questions with Short Answers

1.	Reabsorption	10.	Glomerulus
2.	Alveoli	11.	Lower
3.	Capillaries	12.	Systemic
4.	Nephrons	13.	Hemoglobin
5.	Arteries	14.	Plasma
6.	Liver	15.	Trachea
7.	Diaphragm	16.	B-cells
8.	Duodenum	17.	Macrophages
9.	Villi	18.	Antigens

Label/Diagram/Explain

1. a. The oxygen concentration in the air at high elevations is lower than at lower elevations. As a response, the body manufactures more red blood cells and additional hemoglobin to allow for an increased ability to extract oxygen from the atmosphere and enable transportation of sufficient oxygen in the blood.

   b. The pain, coal dust, or tars from the smoke build up a deposit in the alveoli. This reduces the surface area available for the diffusion of gases and the rate of diffusion of oxygen into the blood, causing shortness of breath. In addition, these activities might also lead to emphysema (a reduction in the surface area of the lungs) by causing destruction of some of the alveoli.

   c. Carbonic anhydrase is the enzyme in red blood cells responsible for converting carbon dioxide molecules into bicarbonate ions. A major portion of the carbon dioxide is carried as bicarbonate ions dissolved in the blood plasma. If carbonic anhydrase were not present, less carbon dioxide could be carried in the blood, and would build up to form carbonic acid. This would lower the pH of the blood and lead to hyperventilation. The person would need to reduce their activity levels considerably.

2. See text figures 18.3 and 18.12.

Multiple Choice Questions

1.	C	6.	C	11.	C
2.	D	7.	A	12.	A
3.	D	8.	A	13.	D
4.	C	9.	C	14.	B
5.	B	10.	D	15.	B

CHAPTER 19 - NUTRITION: FOOD AND DIET

Questions with Short Answers

1. Nutrients
2. Essential
3. Minerals
4. Milk and other dairy products
5. Obese
6. Infancy
7. Diet
8. Kilocalories
9. Water
10. Fiber
11. Fruits and vegetables
12. Calcium
13. Anorexia nervosa or Bulimia
14. Deficiency
15. Protein
16. Electrolytes

Label/Diagram/Explain

1. This diet will work to cause weight loss as long as energy expenditures are greater than 1000 kcal per day. However, an adequate diet considers more than energy. A person on a carbohydrate-only diet can not construct new proteins or fats, since this diet lacks essential amino acids and fatty acid. This could lead to protein deficiency disease. In addition, the lack of vitamins and minerals could also cause other deficiency diseases. Most people could control weight by simply eating a well-balanced diet that provides fewer kcals and also be exercising more.
2. See text figure 19.2.

Multiple Choice Questions

1.	B	6.	A	11.	C
2.	D	7.	C	12.	D
3.	A	8.	A	13.	B
4.	B	9.	C	14.	D
5.	D	10.	A	15.	D

CHAPTER 20 - THE BODY'S CONTROL MECHANISMS

Questions with Short Answers

1. Motor unit
2. Myosin
3. Hormones
4. Stimulus
5. Axon
6. Cones
7. Glands
8. Central
9. Endocrine
10. Sound
11. Nerve impulse
12. Perception
13. Synapse
14. Retina
15. Acetylcholine
16. Endocrine
17. Sodium

Label/Diagram/Explain

See text figure 20.3.

See text figures 20.2 and 20.4. When a nerve cell is stimulated, a small portion of the cell membrane depolarizes as sodium ions flow into the cell through the membrane. This encourages the depolarization of an adjacent portion of the membrane, and it depolarizes a short time later. In this way a wave of depolarization passes down the length of the nerve cell. Shortly after a portion of the membrane is depolarized, the ionic balance is reestablished, it is repolarized and is ready to be stimulated again.

See text figure 20.5. When a nerve impulse reaches the end of an axon, it releases a neurotransmitter (acetylcholine is a common neurotransmitter) into the synapse. When acetylcholine is released into the synapse, the molecules diffuse across the synapse and bind to the receptors on the dendrite, initiating an impulse in the next neuron. Cholinesterase is an enzyme that destroys acetylcholine, preventing continuous stimulation of the dendrite.

Multiple Choice Questions

1.	C	6.	C	11.	B
2.	A	7.	C	12.	C
3.	D	8.	A	13.	A
4.	A	9.	B	14.	B
5.	B	10.	C	15.	D

CHAPTER 21 - HUMAN REPRODUCTION, SEX, AND SEXUALITY

Questions with Short Answers

1. X and Y
2. Autosomes
3. X
4. Differentiation
5. Hormones
6. Puberty
7. Secondary
8. Follicle-stimulating
9. Menstrual
10. Ovulation
11. Testosterone
12. Spermatogenesis
13. Polar
14. Ejaculation
15. Zygote
16. Placenta

Label/Diagram/Explain

See text figure 21.6.

1.	B	6.	A	11.	A
2.	C	7.	A	12.	C
3.	A	8.	C	13.	A
4.	D	9.	D		
5.	A	10.	A		

CHAPTER 22 - THE ORIGIN OF LIFE AND EVOLUTION OF CELLS

Questions with Short Answers

1.	Biogenesis	10.	Mitochondria
2.	Pasteur	11.	RNA
3.	Reducing	12.	Ozone
4.	Biogenesis	13.	Microorganisms
5.	Heterotrophs	14.	Oxidizing
6.	Autotrophs	15.	Archaea
7.	Eukaryotes	16.	4.5 billion
8.	Reducing	17.	Carl Woese
9.	Sun		

Label/Diagram/Explain

See text figure 22.8.

Multiple Choice Questions

1.	D	6.	B	11.	B
2.	B	7.	C	12.	B
3.	C	8.	B	13.	C
4.	A	9.	D		
5.	A	10.	A		

CHAPTER 23 - THE CLASSIFICATION AND EVOLUTION OF ORGANISMS

Questions with Short Answers

1.	*pyogenes*	9.	Gametophyte
2.	Mycetae	10.	Animals
3.	Linnaeus	11.	Virus
4.	Larval	12.	Host
5.	Binary fission	13.	Phylum
6.	Prokaryotae	14.	Saprophytes
7.	Protista	15.	Phylogeny
8.	Fungi	16.	Class

Label/Diagram/Explain

Your answer to this question will vary depending on your bio-geographic location and the organisms with which your are familiar.

1.	B	7.	C	12.	D
2.	A	8.	D	13.	A
3.	D	9.	A	14.	C
4.	C	10.	D	15.	C
5.	D	11.	C	16.	C
6.	D				

CHAPTER 24 - PROKARYOTAE, PROTISTA, AND MYCETAE

Questions with Short Answers

1.	Colony	9.	Bloom
2.	Nitrogen-fixing	10.	Phytoplankton
3.	Pathogen	11.	Locomotion
4.	*Clostridium*	12.	Mycorrhiza
5.	Algae	13.	Lichen
6.	Benthic	14.	Mycotoxins
7.	30-50%	15.	Extreomohiles (Archaea)
8.	Silicon dioxide		

Label/Diagram/Explain

Prokaryotae Benefits:

Bacteria - Associated with decay and decomposition of organic material. Source of chemicals in many foods (alcohol, cheese) and also antibiotics. Fix nitrogen. Photosynthesize.

Protista Benefits:

Protozoans - Associated with the decay and decomposition of organic material. A basic unit in the food chain serving as food for larger organisms.

Algae - Photosynthesize. Source of carrageenin, agar, alginates, and other important organic molecules.

Mycetae Benefits:

Fungi - Associated with the decay and decomposition of organic material. A basic unit in the food chain serving as food for larger organisms. Source of chemicals in many foods (alcohol, cheese) and also antibiotics. Some fungi are edible (ex. Mushrooms).

Multiple Choice Questions

1.	A	7.	C	12.	A
2.	C	8.	A	13.	C
3.	A	9.	D	14.	C
4.	D	10.	C	15.	D
5.	D	11.	D	16.	D
6.	B				

CHAPTER 25 - PLANTAE

Questions with Short Answers

1.	Bryophytes	8.	Pollen
2.	Gametophyte	9.	Theophrastus
3.	Vascular	10.	Wood
4.	Xylem	11.	Fruit
5.	Root hairs	12.	Cotyledons
6.	Sporophyte	13.	Perfect
7.	Cones		

Label/Diagram/Explain

See text figure 25.10 and 25.19.

Multiple Choice Questions

1.	B	6.	C	11.	C
2.	D	7.	A	12.	B
3.	C	8.	A	13.	C
4.	B	9.	C	14.	B
5.	A	10.	C	15.	B

CHAPTER 26 - ANIMALIA

Questions with Short Answers

1.	Inside	9.	Parasitic worms
2.	Poikilotherm	10.	Mollusks
3.	Coelom	11.	Starfish
4.	Sessile	12.	Pelagic
5.	Radial	13.	Cartilage
6.	Vertebrae	14.	Arthropods
7.	Invertebrates	15.	Hibernation
8.	Medusa		

Label/Diagram/Explain

Arthropods: Jointed legs * - Chitinous exoskeleton, compound eyes - Three body parts -Fertilization usually internal* - Striated muscle fibers.

Amphibians: Usually moist skin, no scales -Aquatic larvae and terrestrial adults - Usually two pairs of legs* - Lungs present in adults * - Usually external fertilization.

Reptiles: Scales present - Internal fertilization* - Amniotic egg* - Imperfectly formed four-chambered heart* - Respiration by lungs* - Poikilothermous.

Birds: Body covered with feathers* - Forelimbs usually adapted for flight* - Complete four-chambered heart* - Beak present, no teeth - Amniotic egg* - Homothermous*.

Mammals: Body covered with hair* - Mammary glands present - Male copulatory organ* - Internal fertilization* - Placenta present * - Limbs usually have five digits - Complete four-chambered heart*

Multiple Choice Questions

1.	B	6.	D	11.	C
2.	C	7.	A	12.	D
3.	A	8.	D	13.	A
4.	D	9.	D	14.	A
5.	D	10.	D	15.	A